THE

ALGEBRA

OF

MOHAMMED BEN MUSA.

———

THE

·ALGEBRA·

OF

MOHAMMED BEN MUSA.

EDITED AND TRANSLATED

BY

FREDERIC ROSEN.

LONDON:

PRINTED FOR THE ORIENTAL TRANSLATION FUND:

AND SOLD BY

J. MURRAY, ALBEMARLE STREET;

PARBURY, ALLEN, & CO., LEADENHALL STREET;

THACKER & CO., CALCUTTA; TREUTTEL & WUERTZ, PARIS;

AND E. FLEISCHER, LEIPZIG.

1831.

PRINTED BY
J. L. COX, GREAT QUEEN STREET,
LONDON.

PREFACE.

In the study of history, the attention of the observer is drawn by a peculiar charm towards those epochs, at which nations, after having secured their independence externally, strive to obtain an inward guarantee for their power, by acquiring eminence as great in science and in every art of peace as they have already attained in the field of war. Such an epoch was, in the history of the Arabs, that of the Caliphs AL MANSUR, HARUN AL RASHID, and AL MAMUN, the illustrious contemporaries of CHARLEMAGNE; to the glory of which era, in the volume now offered to the public, a new monument is endeavoured to be raised.

ABU ABDALLAH MOHAMMED BEN MUSA, of Khowarezm, who it appears, from his preface, wrote this Treatise at the command of the Caliph AL MAMUN, was for a long time considered as the original inventor of Algebra. "*Hæc ars olim a* MAHOMETE, Mosis *Arabis filio, initium sumsit: etenim hujus rci locuples testis* LEO-

NARDUS PISANUS." Such are the words with which HIERONYMUS CARDANUS commences his *Ars Magna*, in which he frequently refers to the work here translated, in a manner to leave no doubt of its identity.

That he was not the inventor of the Art, is now well established; but that he was the first Mohammedan who wrote upon it, is to be found asserted in several Oriental writers. HAJI KHALFA, in his bibliographical work, cites the initial words of the treatise now before us,* and

---

* I am indebted to the kindness of my friend Mr. GUSTAV FLUEGEL of Dresden, for a most interesting extract from this part of HAJI KHALFA's work. Complete manuscript copies of the كشف الظنون are very scarce. The only two which I have hitherto had an opportunity of examining (the one bought in Egypt by Dr. EHRENBERG, and now deposited in the Royal Library at Berlin—the other among RICH's collection in the British Museum) are only abridgments of the original compilation, in which the quotation of the initial words of each work is generally omitted. The prospect of an edition and Latin translation of the complete original work, to be published by Mr. FLUEGEL, under the auspices of the Oriental Translation Committee, must under such circumstances be most gratifying to all friends of Asiatic literature.

states, in two distinct passages, that its author, MOHAMMED BEN MUSA, was the first Mussulman who had ever written on the solution of problems by the rules of completion and reduction. Two marginal notes in the Oxford manuscript—from which the text of the present edition is taken—and an anonymous Arabic writer, whose *Bibliotheca Philosophorum* is frequently quoted by CASIRI,* likewise maintain that this production of MOHAMMED BEN MUSA was the first work written on the subject† by a Mohammedan.

---

* تاريخ الحكماء, written in the twelfth century. CASIRI *Bibliotheca Arabica Escurialensis*, T. I. 426. 428.

† The first of these marginal notes stands at the top of the first page of the manuscript, and reads thus : هذا اول كتاب وضع في الجبر والمقابلة في الاسلام ولهذا ذكر فيه من كل فن طرفا ليفيد الاصول في الجبر والمقابلة " This is the first book written on (the art of calculating by) completion and reduction by a Mohammedan: on this account the author has introduced into it rules of various kinds, in order to render useful the very rudiments of Algebra." The other scholium stands farther on: it is the same to which I have referred in my notes to the Arabic text, p. 177.

From the manner in which our author, in his preface, speaks of the task he had undertaken, we cannot infer that he claimed to be the inventor. He says that the Caliph AL MAMUN encouraged him to write a *popular* work on Algebra: an expression which would seem to imply that other treatises were then already extant. From a formula for finding the circumference of the circle, which occurs in the work itself (Text p. 51, Transl. p. 72), I have, in a note, drawn the conclusion, that part of the information comprised in this volume was derived from an Indian source; a conjecture which is supported by the direct assertion of the author of the *Bibliotheca Philosophorum* quoted by CASIRI (I.426, 428). That MOHAMMED BEN MUSA was conversant with Hindu science, is further evident from the fact* that he abridged, at AL MAMUN's request—but before the accession of that prince to the caliphat—the *Sindhind*, or

---

* Related by EBN AL ADAMI in the preface to his astronomical tables. CASIRI, I. 427, 428. COLEBROOKE, Dissertation, &c. p. lxiv. lxxii.

astronomical tables, translated by MOHAM-
MED BEN IBRAHIM AL FAZARI from the
work of an Indian astronomer who visited the
court of ALMANSUR in the 156th year of the
Hejira (A.D. 773).

The science as taught by MOHAMMED BEN
MUSA, in the treatise now before us, does not
extend beyond quadratic equations, including
problems with an affected square. These he
solves by the same rules which are followed by
DIOPHANTUS*, and which are taught, though
less comprehensively, by the Hindu mathemati-
cians†. That he should have borrowed from
DIOPHANTUS is not at all probable ; for it does
not appear that the Arabs had any knowledge
of DIOPHANTUS' work before the middle of the
fourth century after the Hejira, when ABU'L-
WAFA BUZJANI rendered it into Arabic‡.   It

---

* See DIOPHANTUS, Introd. § II. and Book iv. problems 32 and 33.

† *Lilavati*, p. 29, *Vijaganita*, p. 347, of Mr. COLE-
BROOKE's translation.

‡ CASIRI *Bibl. Arab. Escur.* I. 433. COLEBROOKE's
Dissertation, &c. p. lxxii.

is far more probable that the Arabs received their first knowledge of Algebra from the Hindus, who furnished them with the decimal notation of numerals, and with various important points of mathematical and astronomical information.

But under whatever obligation our author may be to the Hindus, as to the subject matter of his performance, he seems to have been independent of them in the manner of digesting and treating it : at least the method which he follows in expounding his rules, as well as in showing their application, differs considerably from that of the Hindu mathematical writers. BHASKARÁ and BRAHMAGUPTA give dogmatical precepts, unsupported by argument, which, even by the metrical form in which they are expressed, seem to address themselves rather to the memory than to the reasoning faculty of the learner: MOHAMMED gives his rules in simple prose, and establishes their accuracy by geometrical illustrations. The Hindus give comparatively few examples, and are fond of investing the statement of their problems in rhetorical pomp: the Arab, on the contrary, is remarkably rich in examples, but he introduces them with the same perspicuous simplicity of style which distinguishes his rules. In solving their problems, the Hindus are satisfied with pointing at the result, and at the principal intermediate steps which lead to it: the Arab shows the working of each example at full length, keeping his view constantly fixed upon the two sides of the equation, as upon the two scales of a balance, and showing how any alteration in one side is counterpoised by a corresponding change in the other.

Besides the few facts which have already been mentioned in the course of this preface, little or nothing is known of our Author's life. He lived and wrote under the caliphat of AL MAMUN, and must therefore be distinguished from ABU JAFAR MOHAMMED BEN MUSA*,

---

* The father of the latter, MUSA BEN SHAKER, whose native country I do not find recorded, had been a robber or bandit in the earlier part of his life, but had afterwards found means to attach himself to the court of the Caliph AL-MAMUN; who, after MUSA's death, took care of

likewise a mathematician and astronomer, who flourished under the Caliph AL MOTADED (who reigned A.H. 279-289, A.D. 892-902).

---

the education of his three sons, MOHAMMED, AHMED, and AL HASSAN. (ABILFARAGII *Histor. Dyn.* p. 280. CASIRI, I. 386. 418). Each of the sons subsequently distinguished himself in mathematics and astronomy. We learn from ABULFARAJ (*l. c.* p. 281) and from EBN KHALLIKAN (art. ثابت بن قرة) that THABET BEN KORRAH, the well-known translator of the Almagest, was indebted to MOHAMMED for his introduction to AL MOTADED, and the men of science at the court of that caliph. EBN KHALLI-KAN's words are : فخرج من حرّان ونزل كَفَرتُوثَا واقام بها مدة الي ان قدم محمد بن موسي من بلاد الروم راجعا الي بغداد فاجتمع به فرأه فاضلا فصيحا فاستصحبه الي بغداد وانزله في داره ووصله بالخليفة فادخله في جملة المنجمين "(THABET BEN KORRAH) left Harran, and established himself at Kafratutha, where he remained till MOHAMMED BEN MUSA arrived there, on his return from the Greek dominions to Bagdad. The latter became acquainted with THABET and on seeing his skill and sagacity, invited THABET to accompany him to Bagdad, where MOHAMMED made him lodge at his own house, introduced him to the Caliph, and procured him an appointment in the body of astronomers." EBN KHALLIKAN here speaks of MOHAMMED BEN MUSA as of a well-known individual: he has however devoted no special article to an account of his life. It is possible

The manuscript from whence the text of the present edition is taken—and which is the only copy the existence of which I have as yet been able to trace—is preserved in the Bodleian collection at Oxford. It is, together with three other treatises on Arithmetic and Algebra, contained in the volume marked CMXVIII. *Hunt.* 214, *fol.*, and bears the date of the transcription A.H. 743 (A. D. 1342). It is written in a plain and legible hand, but unfortunately destitute of most of the diacritical points : a deficiency which has often been very sensibly felt; for though the nature of the subject matter can but seldom leave a doubt as to the general import of a sentence, yet the true reading of some passages, and the precise interpretation of others, remain involved in obscurity. Besides, there occur several omissions of words, and even of entire sentences; and also instances of words or short passages writ-

---

that the tour into the provinces of the Eastern Roman Empire here mentioned, was undertaken in search of some ancient Greek works on mathematics or astronomy.

ten twice over, or words foreign to the sense introduced into the text. In printing the Arabic part, I have included in brackets many of those words which I found in the manuscript, the genuineness of which I suspected, and also such as I inserted from my own conjecture, to supply an apparent hiatus.

The margin of the manuscript is partially filled with *scholia* in a very small and almost illegible character, a few specimens of which will be found in the notes appended to my translation. Some of them are marked as being extracted from a commentary (شرح) by AL MOZAIHAFI*, probably the same author, whose full name is JEMALEDDIN ABU ABDALLAH MOHAMMED BEN OMAR AL JAZA'I† AL MOZAIHAFI, and whose "Introduction to Arithmetic," (مقدمة في الحساب) is contained in the same volume with MOHAMMED's work in the Bodleian library.

Numerals are in the text of the work always

---

* Wherever I have met with this name, it is written without the diacritical points المرحفى, and my pronunciation rests on mere conjecture.

† المحراعى (?)

expressed by words: figures are only used in some of the diagrams, and in a few marginal notes.

The work had been only briefly mentioned in URIS' catalogue of the Bodleian manuscripts. Mr. H. T. COLEBROOKE first introduced it to more general notice, by inserting a full account of it, with an English translation of the directions for the solution of equations, simple and compound, into the notes of the " *Dissertation*" prefixed to his invaluable work, " *Algebra, with Arithmetic and Mensuration, from the Sanscrit of Brahmegupta and Bhascara.*" (London, 1817, 4to. pages lxxv-lxxix.)

The account of the work given by Mr. COLE-BROOKE excited the attention of a highly distinguished friend of mathematical science, who encouraged me to undertake an edition and translation of the whole : and who has taken the kindest interest in the execution of my task. He has with great patience and care revised and corrected my translation, and has furnished the commentary, subjoined to the text, in the form of common algebraic notation. But my obligations to him are not confined to this only ; for his luminous advice has enabled me to overcome many difficulties, which, to my own limited proficiency in mathematics, would have been almost insurmountable.

In some notes on the Arabic text which are appended to my translation, I have endeavoured, not so much to elucidate, as to point out for further enquiry, a few circumstances connected with the history of Algebra. The comparisons drawn between the Algebra of the Arabs and that of the early Italian writers might perhaps have been more numerous and more detailed; but my enquiry was here restricted by the want of some important works. MONTUCLA, COSSALI, HUTTON, and the Basil edition of CARDANUS' *Ars magna*, were the only sources which I had the opportunity of consulting.

THE AUTHOR'S PREFACE.

In the Name of God, gracious and merciful!

This work was written by Mohammed ben Musa, of Khowarezm. He commences it thus:

Praised be God for his bounty towards those who deserve it by their virtuous acts: in performing which, as by him prescribed to his adoring creatures, we express our thanks, and render ourselves worthy of the continuance (of his mercy), and preserve ourselves from change: acknowledging his might, bending before his power, and revering his greatness! He sent Mohammed (on whom may the blessing of God repose!) with the mission of a prophet, long after any messenger from above had appeared, when justice had fallen into neglect, and when the true way of life was sought for in vain. Through him he cured of blindness, and saved through him from perdition, and increased

B

through him what before was small, and collected
through him what before was scattered. Praised be
God our Lord! and may his glory increase, and may
all his names be hallowed—besides whom there is no
God; and may his benediction rest on MOHAMMED
the Prophet and on his descendants!

The learned in times which have passed away, and
among nations which have ceased to exist, were constantly employed in writing books on the several de-
partments of science and on the various branches of
knowledge, bearing in mind those that were to come
after them, and hoping for a reward proportionate to
their ability, and trusting that their endeavours would
meet with acknowledgment, attention, and remem-
brance—content as they were even with a small degree
of praise; small, if compared with the pains which they
had undergone, and the difficulties which they had
encountered in revealing the secrets and obscurities of
science.

(2) Some applied themselves to obtain information which
was not known before them, and left it to posterity;
others commented upon the difficulties in the works
left by their predecessors, and defined the best method
(of study), or rendered the access (to science) easier or

placed it more within reach; others again discovered mistakes in preceding works, and arranged that which was confused, or adjusted what was irregular, and corrected the faults of their fellow-labourers, without arrogance towards them, or taking pride in what they did themselves.

That fondness for science, by which God has distinguished the IMAM AL MAMUN, the Commander of the Faithful (besides the caliphat which He has vouchsafed unto him by lawful succession, in the robe of which He has invested him, and with the honours of which He has adorned him), that affability and condescension which he shows to the learned, that promptitude with which he protects and supports them in the elucidation of obscurities and in the removal of difficulties, —has encouraged me to compose a short work on Calculating by (the rules of) Completion and Reduction, confining it to what is easiest and most useful in arithmetic, such as men constantly require in cases of inheritance, legacies, partition, law-suits, and trade, and in all their dealings with one another, or where the measuring of lands, the digging of canals, geometrical computation, and other objects of various sorts and kinds are concerned—relying on the goodness of my intention therein, and hoping that the learned will reward it, by obtaining (for me) through their prayers the excellence of the Divine mercy: in requital of which, may the choicest blessings and the abundant bounty of God be theirs! My confidence rests with God, in this as in every thing, and in Him I put my trust. He is the Lord of the Sublime Throne. May His blessing descend upon all the prophets and heavenly messengers!

MOHAMMED BEN MUSA'S

COMPENDIUM

ON CALCULATING BY

COMPLETION AND REDUCTION.

———◆———

When I considered what people generally want in (3) calculating, I found that it always is a number.

I also observed that every number is composed of units, and that any number may be divided into units.

Moreover, I found that every number, which may be expressed from one to ten, surpasses the preceding by one unit: afterwards the ten is doubled or tripled, just as before the units were: thus arise twenty, thirty, &c., until a hundred; then the hundred is doubled and tripled in the same manner as the units and the tens, up to a thousand; then the thousand can be thus repeated at any complex number; and so forth to the utmost limit of numeration.

I observed that the numbers which are required in calculating by Completion and Reduction are of three kinds, namely, roots, squares, and simple numbers relative to neither root nor square.

A root is any quantity which is to be multiplied by itself, consisting of units, or numbers ascending, or fractions descending.*

A square is the whole amount of the root multiplied by itself.

A simple number is any number which may be pronounced without reference to root or square.

A number belonging to one of these three classes may be equal to a number of another class; you may say, for instance, " squares are equal to roots," or "squares are equal to numbers," or "roots are equal to numbers."†

(4)   Of the case in which *squares are equal to roots*, this is an example. " A square is equal to five roots of the same;"‡ the root of the square is five, and the square is twenty-five, which is equal to five times its root.

So you say, " one third of the square is equal to four roots;"§ then the whole square is equal to twelve roots; that is a hundred and forty-four; and its root is twelve.

Or you say, " five squares are equal to ten roots;"‖ then one square is equal to two roots; the root of the square is two, and its square is four.

---

* By the word root, is meant the simple power of the unknown quantity.

$$† \; cx^2 = bx \qquad cx^2 = a \qquad bx = a$$
$$‡ \; x^2 = 5x \qquad \therefore \; x = 5$$
$$§ \; \frac{x^2}{3} = 4x \qquad \therefore \; x^2 = 12x \qquad \therefore \; x = 12$$
$$‖ \; 5x^2 = 10x \qquad \therefore \; x^2 = 2x \qquad \therefore \; x = 2$$

In this manner, whether the squares be many or few, (*i. e.* multiplied or divided by any number), they are reduced to a single square ; and the same is done with the roots, which are their equivalents ; that is to say, they are reduced in the same proportion as the squares.

As to the case in which *squares are equal to numbers*; for instance, you say, " a square is equal to nine ;"[*] then this is a square, and its root is three. Or " five squares are equal to eighty ;"[†] then one square is equal to one-fifth of eighty, which is sixteen. Or " the half of the square is equal to eighteen ;"[‡] then the square is thirty-six, and its root is six.

Thus, all squares, multiples, and sub-multiples of them, are reduced to a single square. If there be only part of a square, you add thereto, until there is a whole square ; you do the same with the equivalent in numbers.

As to the case in which *roots are equal to numbers*; for instance, " one root equals three in number ; "[§] then the root is three, and its square nine. Or " four roots (5) are equal to twenty ;"[‖] then one root is equal to five, and the square to be formed of it is twenty-five. Or " half the root is equal to ten ; "[¶] then the

---

[*] $x^2 = 9$ $x = 3$

[†] $5x^2 = 80 \therefore x^2 = \frac{80}{5} = 16$

[‡] $\frac{x^2}{2} = 18 \therefore x^2 = 36 \therefore x = 6$

[§] $x = 3$

[‖] $4x = 20$ $\therefore x = 5$

[¶] $\frac{x}{2} = 10$ $\therefore x = 20$

whole root is equal to twenty, and the square which is formed of it is four hundred.

I found that these three kinds; namely, roots, squares, and numbers, may be combined together, and thus three compound species arise ;* that is, " squares and roots equal to numbers ;" " squares and numbers equal to roots ;" "roots and numbers equal to squares."

Roots and Squares are equal to Numbers ;† for instance, " one square, and ten roots of the same, amount to thirty-nine dirhems ;" that is to say, what must be the square which, when increased by ten of its own roots, amounts to thirty-nine? The solution is this : you halve the number‡ of the roots, which in the present instance yields five. This you multiply by itself; the product is twenty-five. Add this to thirty-nine; the sum is sixty-four. Now take the root of this, which is eight, and subtract from it half the number of the roots, which is five; the remainder is three. This is the root of the square which you sought for; the square itself is nine.

---

* The three cases considered are,

1st. $cx^2 + bx = a$

2d. $cx^2 + a = bx$

3d. $cx^2 = bx + a$

† 1st case: $cx^2 + bx = a$

Example $x^2 + 10x = 39$

$x = \surd[(\frac{1}{2})^2 + 39] - \frac{10}{2}$

$= \surd 64 - 5$

$= 8 - 5 = 3$

‡ *i. e.* the coefficient.

The solution is the same when two squares or three, or more or less be specified;* you reduce them to one single square, and in the same proportion you reduce also the roots and simple numbers which are connected therewith.

For instance, " two squares and ten roots are equal to forty-eight dirhems;"† that is to say, what must be (6) the amount of two squares which, when summed up and added to ten times the root of one of them, make up a sum of forty-eight dirhems? You must at first reduce the two squares to one; and you know that one square of the two is the moiety of both. Then reduce every thing mentioned in the statement to its half, and it will be the same as if the question had been, a square and five roots of the same are equal to twenty-four dirhems; or, what must be the amount of a square which, when added to five times its root, is equal to twenty-four dirhems? Now halve the number of the roots; the moiety is two and a half. Multiply that by itself; the product is six and a quarter. Add this to twenty-four; the sum is thirty dirhems and a quarter. Take the root of this; it is five and a half. Subtract from this the moiety of the number of the roots, that is two and a half; the

---

* $cx^2 + bx = a$ is to be reduced to the form $x^2 + \frac{b}{c}x = \frac{a}{c}$

$$† \quad 2x^2 + 10x = 48$$
$$x^2 + 5x = 24$$
$$x = \sqrt{[(\tfrac{5}{2})^2 + 24]} - \tfrac{5}{2}$$
$$= \sqrt{[6\tfrac{1}{4} + 24]} - 2\tfrac{1}{4}$$
$$= \quad 5\tfrac{1}{2} - 2\tfrac{1}{2} = 3$$

c

remainder is three. This is the root of the square, and the square itself is nine.

The proceeding will be the same if the instance be, " half of a square and five roots are equal to twenty-eight dirhems;"* that is to say, what must be the amount of a square, the moiety of which, when added to the equivalent of five of its roots, is equal to twenty-eight dirhems? Your first business must be to complete your square, so that it amounts to one whole square. This you effect by doubling it. Therefore double it, and double also that which is added to it, as well as what is equal to it. Then you have a square and ten roots, equal to fifty-six dirhems. Now halve the roots; the moiety is five. Multiply this by itself; the product is twenty-five. Add this to fifty-six; the sum is eighty-one. Extract the root of this; it is nine. Subtract from this the moiety of the number of roots, which is five; the remainder is four. This is the root of the square which you sought for; the square is sixteen, and half the square eight.

(7) square eight.

Proceed in this manner, whenever you meet with squares and roots that are equal to simple numbers: for it will always answer.

$$* \frac{x^2}{2} + 5x = 28$$
$$x^2 + 10x = 56$$
$$x = \sqrt{\left[\left(\tfrac{10}{2}\right)^2 + 56\right]} - \tfrac{10}{2}$$
$$= \sqrt{25 + 56} - 5$$
$$= \sqrt{81} - 5$$
$$= 9 - 5 = 4$$

Squares and Numbers are equal to Roots; for instance, "a square and twenty-one in numbers are equal to ten roots of the same square." That is to say, what must be the amount of a square, which, when twenty-one dirhems are added to it, becomes equal to the equivalent of ten roots of that square? Solution: Halve the number of the roots; the moiety is five. Multiply this by itself; the product is twenty-five. Subtract from this the twenty-one which are connected with the square; the remainder is four. Extract its root; it is two. Subtract this from the moiety of the roots, which is five; the remainder is three. This is the root of the square which you required, and the square is nine. Or you may add the root to the moiety of the roots; the sum is seven; this is the root of the square which you sought for, and the square itself is forty-nine.

When you meet with an instance which refers you to this case, try its solution by addition, and if that do not serve, then subtraction certainly will. For in this case both addition and subtraction may be employed, which will not answer in any other of the three cases in which

---

* 2d case. $cx^2 + a = bx$

  Example. $x^2 + 21 = 10x$

  $$x = \tfrac{10}{2} \pm \sqrt{[(\tfrac{10}{2})^2 - 21]}$$
  $$= 5 \pm \sqrt{25 - 21}$$
  $$= 5 \pm \sqrt{4}$$
  $$= 5 \pm 2$$

the number of the roots must be halved. And know, that, when in a question belonging to this case you have halved the number of the roots and multiplied the moiety by itself, if the product be less than the number of dirhems connected with the square, then the instance is impossible;* but if the product be equal to

(8) the dirhems by themselves, then the root of the square is equal to the moiety of the roots alone, without either addition or subtraction.

In every instance where you have two squares, or more or less, reduce them to one entire square, † as I have explained under the first case.

Roots and Numbers are equal to Squares;‡ for instance, " three roots and four of simple numbers are equal to a square." Solution: Halve the roots; the moiety is one and a half. Multiply this by itself; the product is two and a quarter. Add this to the four; the sum is

---

* If in an equation, of the form $x^2 + a = bx$, $(\frac{b}{2})^2 \angle a$, the case supposed in the equation cannot happen. If $(\frac{b}{2})^2 = a$, then $x = \frac{b}{2}$

† $cx^2 + a = bx$ is to be reduced to $x^2 + \frac{a}{c} = \frac{b}{c}x$

‡ 3d case $cx^2 = bx + a$

Example $x^2 = 3x + 4$

$$x^2 = \sqrt{[(\tfrac{3}{2})^2 + 4]} + \tfrac{3}{2}$$
$$= \sqrt{(1\tfrac{1}{4})^2 + 4} + 1\tfrac{1}{2}$$
$$= \sqrt{2\tfrac{1}{4} + 4} + 1\tfrac{1}{2}$$
$$= \sqrt{6\tfrac{1}{4}} + 1\tfrac{1}{2}$$
$$= 2\tfrac{1}{2} + 1\tfrac{1}{2} = 4$$

six and a quarter. Extract its root; it is two and a
half. Add this to the moiety of the roots, which was
one and a half; the sum is four. This is the root of the
square, and the square is sixteen.

Whenever you meet with a multiple or sub-multiple
of a square, reduce it to one entire square.

These are the six cases which I mentioned in the
introduction to this book. They have now been explained. I have shown that three among them do not
require that the roots be halved, and I have taught
how they must be resolved. As for the other three, in
which halving the roots is necessary, I think it expe-
dient, more accurately, to explain them by separate
chapters, in which a figure will be given for each
case, to point out the reasons for halving.

*Demonstration of the Case: " a Square and ten Roots
are equal to thirty-nine Dirhems."**

The figure to explain this a quadrate, the sides of
which are unknown. It represents the square, the
which, or the root of which, you wish to know. This is
the figure A B, each side of which may be considered
as one of its roots; and if you multiply one of these   (9)
sides by any number, then the amount of that number
may be looked upon as the number of the roots which
are added to the square. Each side of the quadrate
represents the root of the square; and, as in the instance,

---

* Geometrical illustration of the case, $x^2 + 10x = 39$

the roots were connected with the square, we may take one-fourth of ten, that is to say, two and a half, and combine it with each of the four sides of the figure. Thus with the original quadrate A B, four new parallelograms are combined, each having a side of the quadrate as its length, and the number of two and a half as its breadth; they are the parallelograms C, G, T, and K. We have now a quadrate of equal, though unknown sides; but in each of the four corners of which a square piece of two and a half multiplied by two and a half is wanting. In order to compensate for this want and to complete the quadrate, we must add (to that which we have already) four times the square of two and a half, that is, twenty-five. We know (by the statement) that the first figure, namely, the quadrate representing the square, together with the four parallelograms around it, which represent the ten roots, is equal to thirty-nine of numbers. If to this we add twenty-five, which is the equivalent of the four quadrates at the corners of the figure A B, by which the great figure D H is completed, then we know that this together makes sixty-four. One side of this great quadrate is its root, that is, eight. If we subtract twice a fourth of ten, that is five, from eight, as from the two extremities of the side of the great quadrate D H, then the remainder of such a side will be three, and that is the root of the square, or the side of the original figure A B. It must be observed, that we have halved the number of the roots, and added the product of the moiety multiplied by itself to the number

thirty-nine, in order to complete the great figure in its (10)
four corners; because the fourth of any number multiplied by itself, and then by four, is equal to the product
of the moiety of that number multiplied by itself.*
Accordingly, we multiplied only the moiety of the roots
by itself, instead of multiplying its fourth by itself, and
then by four. This is the figure :

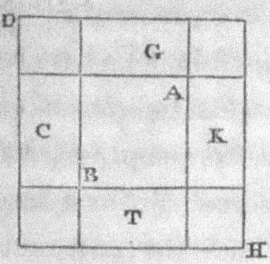

The same may also be explained by another figure.
We proceed from the quadrate A B, which represents
the square. It is our next business to add to it the ten
roots of the same. We halve for this purpose the ten,
so that it becomes five, and construct two quadrangles
on two sides of the quadrate A B, namely, G and D,
the length of each of them being five, as the moiety of
the ten roots, whilst the breadth of each is equal to a
side of the quadrate A B. Then a quadrate remains
opposite the corner of the quadrate A B. This is equal
to five multiplied by five: this five being half of the
number of the roots which we have added to each of the
two sides of the first quadrate. Thus we know that

$$* \quad 4 \times \left(\frac{b}{4}\right)^2 = \left(\frac{b}{2}\right)^2$$

the first quadrate, which is the square, and the two quadrangles on its sides, which are the ten roots, make together thirty-nine. In order to complete the great quadrate, there wants only a square of five multiplied (11) by five, or twenty-five. This we add to thirty-nine, in order to complete the great square S H. The sum is sixty-four. We extract its root, eight, which is one of the sides of the great quadrangle. By subtracting from this the same quantity which we have before added, namely five, we obtain three as the remainder. This is the side of the quadrangle A B, which represents the square; it is the root of this square, and the square itself is nine. This is the figure:—

Demonstration of the Case: " a Square and twenty-one Dirhems are equal to ten Roots."*

We represent the square by a quadrate A D, the length of whose side we do not know. To this we join a parallelogram, the breadth of which is equal to one of the sides of the quadrate A D, such as the side H N. This paralellogram is H B. The length of the two

---

* Geometrical illustration of the case, $x^2 + 21 = 10x$

figures together is equal to the line H C. We know that its length is ten of numbers; for every quadrate has equal sides and angles, and one of its sides multi_ plied by a unit is the root of the quadrate, or multiplied by two it is twice the root of the same. As it is stated, therefore, that a square and twenty-one of numbers are equal to ten roots, we may conclude that the length of the line H C is equal to ten of numbers, since the line C D represents the root of the square. We now divide the line C H into two equal parts at the point G: the line G C is then equal to H G. It is also evident that (12) the line G T is equal to the line C D. At present we add to the line G T, in the same direction, a piece equal to the difference between C G and G T, in order to complete the square. Then the line T K becomes equal to K M, and we have a new quadrate of equal sides and angles, namely, the quadrate M T. We know that the line T K is five; this is consequently the length also of the other sides: the quadrate itself is twenty-five, this being the product of the multiplication of half the number of the roots by themselves, for five times five is twenty-five. We have perceived that the quadrangle H B represents the twenty-one of numbers which were added to the quadrate. We have then cut off a piece from the quadrangle H B by the line K T (which is one of the sides of the quadrate M T), so that only the part T A remains. At present we take from the line K M the piece K L, which is equal to G K; it then appears that the line T G is equal to M L; moreover, the line K L, which has been cut off from K M, is equal to K G; consequently, the quadrangle MR is equal to T A. Thus it is evident that the quadrangle H T, augmented by the quadrangle M R, is equal to the quadrangle H B, which represents the twenty-one. The whole quadrate M T was found to be equal to twenty-five. If we now subtract from this quadrate, M T, the quadrangles H T and M R, which are equal to twenty-one, there remains a small quadrate K R, which represents the difference between twenty-five and twenty-one. This is four; and its root, represented by the line R G, which is equal to G A, is two. If you (13) subtract this number two from the line C G, which is the moiety of the roots, then the remainder is the line A C; that is to say, three, which is the root of the original square. But if you add the number two to the line C G, which is the moiety of the number of the roots, then the sum is seven, represented by the line C R, which is the root to a larger square. However, if you add twenty-one to this square, then the sum will likewise be equal to ten roots of the same square. Here is the figure :—

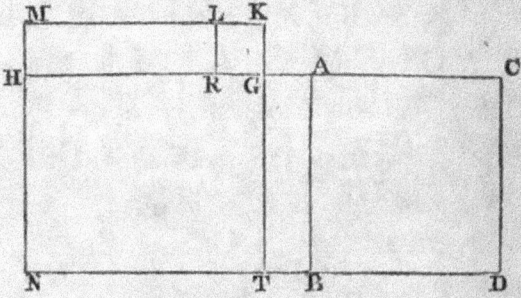

Demonstration of the Case: "three Roots and four of Simple Numbers are equal to a Square." *

Let the square be represented by a quadrangle, the sides of which are unknown to us, though they are equal among themselves, as also the angles. This is the quadrate A D, which comprises the three roots and the four of numbers mentioned in this instance. In every quadrate one of its sides, multiplied by a unit, is its root. We now cut off the quadrangle H D from the quadrate A D, and take one of its sides H C for three, which is the number of the roots. The same is equal to R D. It follows, then, that the quadrangle H B represents the four of numbers which are added to the roots. Now we halve the side C H, which is equal to three roots, at the point G; from this division we construct the square H T, which is the product of half the roots (or one and (14) a half) multiplied by themselves, that is to say, two and a quarter. We add then to the line G T a piece equal to the line A H, namely, the piece T L; accordingly the line G L becomes equal to A G, and the line K N equal to T L. Thus a new quadrangle, with equal sides and angles, arises, namely, the quadrangle G M; and we find that the line A G is equal to M L, and the same line A G is equal to G L. By these means the line C G remains equal to N R, and the line M N equal to T L, and from the quadrangle H B a piece equal to the quadrangle K L is cut off.

---

* Geometrical illustration of the 3d case, $x^2 = 3x + 4$

But we know that the quadrangle A R represents the four of numbers which are added to the three roots. The quadrangle A N and the quadrangle K L are together equal to the quadrangle A R, which represents the four of numbers.

We have seen, also, that the quadrangle G M comprises the product of the moiety of the roots, or of one and a half, multiplied by itself; that is to say two and a quarter, together with the four of numbers, which are represented by the quadrangles A N and K L. There remains now from the side of the great original quadrate A D, which represents the whole square, only the moiety of the roots, that is to say, one and a half, namely, the line G C. If we add this to the line A G, which is the root of the quadrate G M, being equal to two and a half; then this, together with C G, or the moiety of the three roots, namely, one and a half, makes four, which is the line A C, or the root to a square, which is represented by the quadrate A D. Here follows the figure. This it was which we were desirous to explain.

(15)

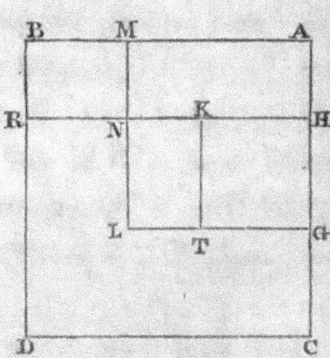

We have observed that every question which requires equation or reduction for its solution, will refer you to one of the six cases which I have proposed in this book. I have now also explained their arguments. Bear them, therefore, in mind.

ON MULTIPLICATION.

I SHALL now teach you how to multiply the unknown numbers, that is to say, the roots, one by the other, if they stand alone, or if numbers are added to them, or if numbers are subtracted from them, or if they are subtracted from numbers; also how to add them one to the other, or how to subtract one from the other.

Whenever one number is to be multiplied by another, the one must be repeated as many times as the other contains units.*

If there are greater numbers combined with units to be added to or subtracted from them, then four multiplications are necessary;† namely, the greater numbers by the greater numbers, the greater numbers by the

---

* If x is to be multiplied by y, x is to be repeated as many times as there are units in y.

† If $x \pm a$ is to be multiplied by $y \pm b$, x is to be multiplied by y, x is to be multiplied by b, a is to be multiplied by y, and a is to be multiplied by b.

units, the units by the greater numbers, and the units by the units.

If the units, combined with the greater numbers, are positive, then the last multiplication is positive; if they are both negative, then the fourth multiplication is likewise positive. But if one of them is positive, and one (16) negative, then the fourth multiplication is negative.*

For instance, " ten and one to be multiplied by ten and two."† Ten times ten is a hundred; once ten is ten positive; twice ten is twenty positive, and once two is two positive; this altogether makes a hundred and thirty-two.

But if the instance is " ten less one, to be multiplied by ten less one,"‡ then ten times ten is a hundred; the

---

* In multiplying $(x \pm a)$ by $(y \pm b)$

$$+a \times +b = +ab$$
$$-a \times -b = +ab$$
$$+a \times -b = -ab$$
$$-a \times +b = -ab$$

† $(10+1) \times (10+2)$
$$= 10 \times 10 \ldots . 100$$
$$+ \ 1 \times 10 \ldots . \ 10$$
$$+ \ 2 \times 10 \ldots . \ 20$$
$$+ \ 1 \times \ 2 \ldots . \ 2$$
$$\overline{}$$
$$+132$$

‡ $(10-1) \ (10-1)$
$$= 10 \times \ 10 .. +100$$
$$- \ 1 \times \ 10 .. - \ 10$$
$$- \ 1 \times \ 10 .. - \ 10$$
$$- \ 1 \times -1 .. + \ \ 1$$
$$\overline{}$$
$$+ \ 81$$

negative one by ten is ten negative; the other negative one by ten is likewise ten negative, so that it becomes eighty: but the negative one by the negative one is one positive, and this makes the result eighty-one.

Or if the instance be "ten and two, to be multipled by ten less one,"* then ten times ten is a hundred, and the negative one by ten is ten negative; the positive two by ten is twenty positive; this together is a hundred and ten; the positive two by the negative one gives two negative. This makes the product a hundred and eight.

I have explained this, that it might serve as an introduction to the multiplication of unknown sums, when numbers are added to them, or when numbers are subtracted from them, or when they are subtracted from numbers.

For instance: "Ten less thing (the signification of thing being root) to be multipled by ten."† You begin by taking ten times ten, which is a hundred; less thing by ten is ten roots negative; the product is therefore a hundred less ten things.

$$* \ (10+2) \times (10-1) =$$
$$10 \times 10 \dots \ 100$$
$$- \ 1 \times 10 \dots \ -10$$
$$+10 \times \ 2 \dots \ +20$$
$$- \ 1 \times \ 2 \dots \ - \ 2$$
$$\overline{}$$
$$108$$

$$† \ (10-x) \times 10 = 10 \times 10 - 10x = 100 - 10x.$$

If the instance be: " ten and thing to be multiplied by ten,"* then you take ten times ten, which is a hundred, and thing by ten is ten things positive; so that the product is a hundred plus ten things.

If the instance be: " ten and thing to be multiplied (17) by itself,"† then ten times ten is a hundred, and ten times thing is ten things; and again, ten times thing is ten things; and thing multiplied by thing is a square positive, so that the whole product is a hundred dirhems and twenty things and one positive square.

If the instance be: " ten minus thing to be multiplied by ten minus thing,"‡ then ten times ten is a hundred; and minus thing by ten is minus ten things; and again, minus thing by ten is minus ten things. But minus thing multiplied by minus thing is a positive square. The product is therefore a hundred and a square, minus twenty things.

In like manner if the following question be proposed to you: " one dirhem minus one-sixth to be multiplied by one dirhem minus one-sixth;"§ that is to say, five-sixths by themselves, the product is five and twenty parts of a dirhem, which is divided into six and thirty parts, or two-thirds and one-sixth of a sixth. Computation: You multiply one dirhem by one dirhem, the

---

*$(10+x) \times 10 = 10 \times 10 + 10x = 100 + 10x$

†$(10+x)(10+x) = 10 \times 10 + 10x + 10x + x^2 = 100 + 20x + x^2$

‡$(10-x) \times (10-x) = 10 \times 10 - 10x - 10x + x^2 = 100 - 20x + x^2$

§$(1-\frac{1}{6}) \times (1-\frac{1}{6}) = 1 - \frac{1}{3} + \frac{1}{6} \times \frac{1}{6} = \frac{2}{3} + \frac{1}{6} \times \frac{1}{6}$; i.e. $\frac{25}{36} = \frac{2}{3} + \frac{1}{6} \times \frac{1}{6}$

product is one dirhem; then one dirhem by minus one-sixth, that is one-sixth negative; then, again, one dirhem by minus one-sixth is one-sixth negative: so far, then, the result is two-thirds of a dirhem: but there is still minus one-sixth to be multiplied by minus one-sixth, which is one-sixth of a sixth positive; the product is, therefore, two-thirds and one sixth of a sixth.

If the instance be, " ten minus thing to be multiplied by ten and thing," then you say,* ten times ten is a hundred; and minus thing by ten is ten things negative; and thing by ten is ten things positive; and minus thing by thing is a square positive; therefore, the product is a hundred dirhems, minus a square.

If the instance be, " ten minus thing to be multiplied by thing,"† then you say, ten multiplied by thing is ten things; and minus thing by thing is a square negative; (18) therefore, the product is ten things minus a square.

If the instance be, " ten and thing to be multiplied by thing less ten,"‡ then you say, thing multiplied by ten is ten things positive; and thing by thing is a square positive; and minus ten by ten is a hundred dirhems negative; and minus ten by thing is ten things negative. You say, therefore, a square minus a hundred dirhems; for, having made the reduction, that is to say, having removed the ten things positive by the ten things

---

* $(10-x)\,(10+x) = 10 \times 10 - 10x + 10x - x^2 = 100 - x^2$

† $(10-x) \times x = 10x - x^2$

‡ $(10+x)\,(x-10) = 10x + x^2 - 100 - 10x = x^2 - 100$

E

negative, there remains a square minus a hundred dirhems.

If the instance be, " ten dirhems and half a thing to be multiplied by half a dirhem, minus five things,"* then you say, half a dirhem by ten is five dirhems positive; and half a dirhem by half a thing is a quarter of thing positive; and minus five things by ten dirhems is fifty roots negative. This altogether makes five dirhems minus forty-nine things and three quarters of thing. After this you multiply five roots negative by half a root positive: it is two squares and a half negative. Therefore, the product is five dirhems, minus two squares and a half, minus forty-nine roots and three quarters of a root.

If the instance be, "ten and thing to be multiplied by thing less ten,"† then this is the same as if it were said thing and ten by thing less ten. You say, therefore, thing multiplied by thing is a square positive; and ten by thing is ten things positive; and minus ten by thing is ten things negative. You now remove the positive by the negative, then there only remains a square. Minus ten multiplied by ten is a hundred, to be subtracted from the square. This, therefore, altogether, is a square less a hundred dirhems.

(19) Whenever a positive and a negative factor concur in

---

*$(10+\frac{x}{2})(\frac{1}{2}-5x)=\frac{10}{2}+\frac{x}{4}-50x-\frac{5}{2}x^2=5-49\frac{3}{4}x-2\frac{1}{2}x^2$

†$(10+x)(x-10)=(x+10)(x-10)=x^2+10x-10x-100=x^2-100$

a multiplication, such as thing positive and minus thing, the last multiplication gives always the negative product. Keep this in memory.

ON ADDITION AND SUBTRACTION.

Know that the root of two hundred minus ten, added to twenty minus the root of two hundred, is just ten.*

The root of two hundred, minus ten, subtracted from twenty minus the root of two hundred, is thirty minus twice the root of two hundred; twice the root of two hundred is equal to the root of eight hundred.†

A hundred and a square minus twenty roots, added to fifty and ten roots minus two squares,‡ is a hundred and fifty, minus a square and minus ten roots.

A hundred and a square, minus twenty roots, diminished by fifty and ten roots minus two squares, is fifty dirhems and three squares minus thirty roots.§

I shall hereafter explain to you the reason of this by a figure, which will be annexed to this chapter.

If you require to double the root of any known or unknown square, (the meaning of its duplication being

---

* $20-\sqrt{200}+(\sqrt{200}-10)=10$

† $20-\sqrt{200}-(\sqrt{200}-10)=30-2\sqrt{200}=30-\sqrt{800}$

‡ $50+10x-2x^2+(100+x^2-20x)=150-10x-x^2$

§ $100+x^2-20x-[50-2x^2+10x]=50+3x^2-30x$

that you multiply it by two) then it will suffice to multiply two by two, and then by the square;* the root of the product is equal to twice the root of the original square.

If you require to take it thrice, you multiply three by three, and then by the square; the root of the product is thrice the root of the original square.

Compute in this manner every multiplication of the roots, whether the multiplication be more or less than two.†

(20) If you require to find the moiety of the root of the square, you need only multiply a half by a half, which is a quarter; and then this by the square: the root of the product will be half the root of the first square.‡

Follow the same rule when you seek for a third, or a quarter of a root, or any larger or smaller quota§ of it, whatever may be the denominator or the numerator.

Examples of this : If you require to double the root of nine,‖ you multiply two by two, and then by nine: this gives thirty-six; take the root of this, it is six, and this is double the root of nine.

---

* $2\sqrt{x^2} = \sqrt{4x^2}$
  $3\sqrt{x^2} = \sqrt{9x^2}$

† $n\sqrt{x^2} = \sqrt{n^2 x^2}$

‡ $\frac{1}{2}\sqrt{x^2} = \sqrt{\frac{x^2}{4}}$

§ $\frac{1}{n}\sqrt{x^2} = \sqrt{\frac{x^2}{n^2}}$

‖ $2\sqrt{9} = \sqrt{4 \times 9} = \sqrt{36} = 6$

In the same manner, if you require to triple the root of nine,* you multiply three by three, and then by nine: the product is eighty-one; take its root, it is nine, which becomes equal to thrice the root of nine.

If you require to have the moiety of the root of nine,† you multiply a half by a half, which gives a quarter, and then this by nine; the result is two and a quarter: take its root; it is one and a half, which is the moiety of the root of nine.

You proceed in this manner with every root, whether positive or negative, and whether known or unknown.

ON DIVISION.

If you will divide the root of nine by the root of four,‡ you begin with dividing nine by four, which gives two and a quarter: the root of this is the number which you require—it is one and a half.

If you will divide the root of four by the root of nine,§ you divide four by nine; it is four-ninths of the unit: the root of this is two divided by three; namely, two-thirds of the unit.

---

* $3\sqrt{9} = \sqrt{9 \times 9} = \sqrt{81} = 9$

† $\frac{1}{2}\sqrt{9} = \sqrt{\frac{9}{4}} = \sqrt{2\frac{1}{4}} = 1\frac{1}{2}$

‡ $\frac{\sqrt{9}}{\sqrt{4}} = \sqrt{\frac{9}{4}} = \sqrt{2\frac{1}{4}} = 1\frac{1}{2}$

§ $\frac{\sqrt{4}}{\sqrt{9}} = \sqrt{\frac{4}{9}} = \frac{2}{3}$

If you wish to divide twice the root of nine by the root of four, or of any other square*, you double the
(21) root of nine in the manner above shown to you in the chapter on Multiplication, and you divide the product by four, or by any number whatever. You perform this in the way above pointed out.

In like manner, if you wish to divide three roots of nine, or more, or one-half or any multiple or sub-multiple of the root of nine, the rule is always the same:† follow it, the result will be right.

If you wish to multiply the root of nine by the root of four,‡ multiply nine by four; this gives thirty-six; take its root, it is six; this is the root of nine, multiplied by the root of four.

Thus, if you wish to multiply the root of five by the root of ten,§ multiply five by ten: the root of the product is what you have required.

If you wish to multiply the root of one-third by the root of a half,‖ you multiply one-third by a half: it is one-sixth: the root of one-sixth is equal to the root of one-third, multiplied by the root of a half.

If you require to multiply twice the root of nine by

$$* \ \frac{2\sqrt{9}}{\sqrt{4}} = \sqrt{\frac{36}{4}} = \sqrt{9} = 3$$

$$† \ \frac{m\sqrt{p^2}}{\sqrt{q^2}} = \sqrt{\frac{m^2 p^2}{q^2}}$$

$$‡ \ \sqrt{4} \times \sqrt{9} = \sqrt{4 \times 9} = \sqrt{36} = 6$$

$$§ \ \sqrt{10} \times \sqrt{5} = \sqrt{5 \times 10} = \sqrt{50}$$

$$‖ \ \sqrt{\tfrac{1}{2}} \times \sqrt{\tfrac{1}{3}} = \sqrt{\tfrac{1}{2} \times \tfrac{1}{3}} = \sqrt{\tfrac{1}{6}}$$

thrice the root of four,* then take twice the root of nine, according to the rule above given, so that you may know the root of what square it is. You do the same with respect to the three roots of four in order to know what must be the square of such a root. You then multiply these two squares, the one by the other, and the root of the product is equal to twice the root of nine, multiplied by thrice the root of four.

You proceed in this manner with all positive or negative roots.

Demonstrations. (22)

The argument for the root of two hundred, minus ten, added to twenty, minus the root of two hundred, may be elucidated by a figure:

Let the line A B represent the root of two hundred; let the part from A to the point C be the ten, then the remainder of the root of two hundred will correspond to the remainder of the line A B, namely to the line C B. Draw now from the point B a line to the point D, to represent twenty; let it, therefore, be twice as long as the line A C, which represents ten; and mark a part of it from the point B to the point H, to be equal to the line A B, which represents the root of two hundred; then the remainder of the twenty will be equal to the part of the line, from the point H to the point D. As

---

* $3\sqrt{4} \times 2\sqrt{9} = \sqrt{9 \times 4} \times \sqrt{4 \times 9} = \sqrt{36 \times 36} = 36$

our object was to add the remainder of the root of two hundred, after the subtraction of ten, that is to say, the line C B, to the line H D, or to twenty, minus the root of two hundred, we cut off from the line B H a piece equal to C B, namely, the line S H. We know already that the line A B, or the root of two hundred, is equal to the line B H, and that the line A C, which represents the ten, is equal to the line S B, as also that the remainder of the line A B, namely, the line C B is equal to the remainder of the line B H, namely, to S H. Let us add, therefore, this piece S H, to the line H D. We have already seen that from the line B D, or twenty, a piece equal to A C, which is ten, was cut off, namely, the piece B S. There remains after this the line S D, which, consequently, is equal to ten. This it was that we intended to elucidate. Here follows the figure.

(23)

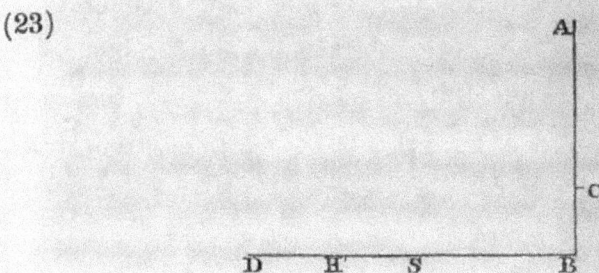

The argument for the root of two hundred, minus ten, to be subtracted from twenty, minus the root of two hundred, is as follows. Let the line A B represent the root of two hundred, and let the part thereof, from A to the point C, signify the ten mentioned in the instance. We draw now from the point B, a line towards the point D, to signify twenty. Then we trace from B to the

point H, the same length as the length of the line which
represents the root of two hundred; that is of the line
A B. We have seen that the line C B is the remainder
from the twenty, after the root of two hundred has been
subtracted. It is our purpose, therefore, to subtract
the line C B from the line H D; and we now draw from
the point B, a line towards the point S, equal in length
to the line A C, which represents the ten. Then the
whole line S D is equal to S B, plus B D, and we perceive that all this added together amounts to thirty.
We now cut off from the line H D, a piece equal to
C B, namely, the line H G; thus we find that the line
G D is the remainder from the line S D, which signifies
thirty. We see also that the line B H is the root of
two hundred and that the line S B and B C is likewise
the root of two hundred. Now the line H G is equal
to C B; therefore the piece subtracted from the line
S D, which represents thirty, is equal to twice the
root of two hundred, or once the root of eight hundred. (24)
This it is that we wished to elucidate.

Here follows the figure:

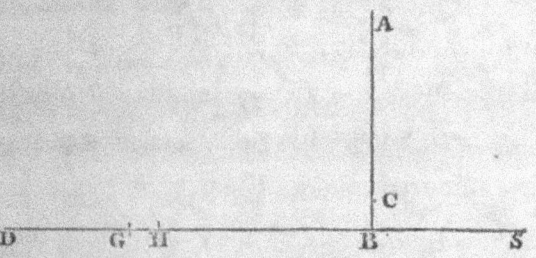

As for the hundred and square minus twenty roots
added to fifty, and ten roots minus two squares, this does

not admit of any figure, because there are three different species, *viz.* squares, and roots, and numbers, and nothing corresponding to them by which they might be represented. We had, indeed, contrived to construct a figure also for this case, but it was not sufficiently clear.

The elucidation by words is very easy. You know that you have a hundred and a square, minus twenty roots. When you add to this fifty and ten roots, it becomes a hundred and fifty and a square, minus ten roots. The reason for these ten negative roots is, that from the twenty negative roots ten positive roots were subtracted by reduction. This being done, there remains a hundred and fifty and a square, minus ten roots. With the hundred a square is connected. If you subtract from this hundred and square the two squares negative connected with fifty, then one square disappears by reason of the other, and the remainder is a hundred and fifty, minus a square, and minus ten roots.

This it was that we wished to explain.

OF THE SIX PROBLEMS.

BEFORE the chapters on computation and the several (25) species thereof, I shall now introduce six problems, as instances of the six cases treated of in the beginning of this work. I have shown that three among these cases, in order to be solved, do not require that the roots be halved, and I have also mentioned that the calculating by completion and reduction must always necessarily lead you to one of these cases. I now subjoin these problems, which will serve to bring the subject nearer to the understanding, to render its comprehension easier, and to make the arguments more perspicuous.

First Problem.

I have divided ten into two portions; I have multiplied the one of the two portions by the other; after this I have multiplied the one of the two by itself, and the product of the multiplication by itself is four times as much as that of one of the portions by the other.*

Computation: Suppose one of the portions to be thing, and the other ten minus thing: you multiply

---

* $x^2 = 4x(10 - x) = 40x - 4x^2$

$5x^2 = 40x$

$x^2 = 8x$

$x = 8; \ (10 - x) = 2$

thing by ten minus thing; it is ten things minus a square. Then multiply it by four, because the instance states "four times as much." The result will be four times the product of one of the parts multiplied by the other. This is forty things minus four squares. After this you multiply thing by thing, that is to say, one of the portions by itself. This is a square, which is equal to forty things minus four squares. Reduce it now by the four squares, and add them to the one square. Then the equation is: forty things are equal to five squares; and one square will be equal to eight roots, that is, sixty-four; the root of this is eight, and this is one of the two portions, namely, that which is to (26) be multiplied by itself. The remainder from the ten is two, and that is the other portion. Thus the question leads you to one of the six cases, namely, that of "squares equal to roots." Remark this.

Second Problem.

I have divided ten into two portions: I have multiplied each of the parts by itself, and afterwards ten by itself: the product of ten by itself is equal to one of the two parts multiplied by itself, and afterwards by two and seven-ninths; or equal to the other multiplied by itself, and afterwards by six and one-fourth.*

$$* \quad 10^2 = x^2 \times 2\tfrac{7}{9}$$
$$100 = x^2 \times \tfrac{25}{9}$$
$$\tfrac{9}{25} \times 100 = x^2$$
$$36 = x^2$$
$$6 = x$$

Computation : Suppose one of the parts to be thing, and the other ten minus thing. You multiply thing by itself, it is a square; then by two and seven-ninths, this makes it two squares and seven-ninths of a square. You afterwards multiply ten by ten; it is a hundred, which much be equal to two squares and seven-ninths of a square. Reduce it to one square, through division by nine twenty-fifths;* this being its fifth and four-fifths of its fifth, take now also the fifth and four-fifths of the fifth of a hundred; this is thirty-six, which is equal to one square. Take its root, it is six. This is one of the two portions; and accordingly the other is four. This question leads you, therefore, to one of the six cases, namely, "squares equal to numbers."

Third Problem.

I have divided ten into two parts. I have afterwards divided the one by the other, and the quotient was four.†

Computation : Suppose one of the two parts to be (27) thing, the other ten minus thing. Then you divide ten minus thing by thing, in order that four may be obtained. You know that if you multiply the quotient by the divisor, the sum which was divided is restored.

---

* $\frac{9}{25} = \frac{1}{5} \times \frac{4}{5} + \frac{1}{5}$

† $\frac{10 - x}{x} = 4$

$10 - x = 4x$

$10 = 5x$

$2 = x$

In the present question the quotient is four and the divisor is thing. Multiply, therefore, four by thing; the result is four things, which are equal to the sum to be divided, which was ten minus thing. You now reduce it by thing, which you add to the four things. Then we have five things equal to ten; therefore one thing is equal to two, and this is one of the two portions. This question refers you to one of the six cases, namely, "roots equal to numbers."

Fourth Problem.

I have multiplied one-third of thing and one dirhem by one-fourth of thing and one dirhem, and the product was twenty.*

Computation: You multiply one-third of thing by one-fourth of thing; it is one-half of a sixth of a square. Further, you multiply one dirhem by one-third of thing, it is one-third of thing; and one dirhem by one-fourth of thing, it is one-fourth of thing; and one dirhem by one dirhem, it is one dirhem. The result of this is: the moiety of one-sixth of a square, and one-third of thing, and one-fourth of thing, and one dirhem, is equal to twenty dirhems. Subtract now the one dirhem from

---

* $\left(\frac{1}{3}x+1\right)\left(\frac{1}{4}x+1\right)=20$

$\frac{x^2}{12}+\frac{1}{3}x+\frac{1}{4}x+1=20$

$\frac{x^2}{12}+\frac{7}{12}x=19$

$x^2+7x=228$

$x=\sqrt{\frac{49}{4}+228}-\frac{7}{2}=12$

these twenty dirhems, there remain nineteen dirhems, equal to the moiety of one-sixth of a square, and one-third of thing, and one-fourth of thing. Now make your square a whole one: you perform this by multiplying all that you have by twelve. Thus you have one square and seven roots, equal to two hundred and twenty-eight dirhems. Halve the number of the roots, and multiply it by itself; it is twelve and one-fourth. Add this to the numbers, that is, to two hundred and twenty-eight; (28) the sum is two hundred and forty and one quarter. Extract the root of this; it is fifteen and a half. Subtract from this the moiety of the roots, that is, three and a half, there remains twelve, which is the square required. This question leads you to one of the cases, namely, " squares and roots equal to numbers."

Fifth Problem.

I have divided ten into two parts; I have then multiplied each of them by itself, and when I had added the products together, the sum was fifty-eight dirhems.*

Computation : Suppose one of the two parts to be thing, and the other ten minus thing. Multiply ten minus thing by itself; it is a hundred and a square minus twenty things. Then multiply thing by thing; it

---

$$* \quad x^2 + (10 - x)^2 = 58$$
$$2x^2 - 20x + 100 = 58$$
$$x^2 - 10x + 50 = 29$$
$$x^2 + 21 = 10x$$
$$x = 5 \pm \sqrt{25 - 21} = 5 \pm 2 = 7 \text{ or } 3$$

is a square. Add both together. The sum is a hundred, plus two squares minus twenty things, which are equal to fifty-eight dirhems. Take now the twenty negative things from the hundred and the two squares, and add them to fifty-eight; then a hundred, plus two squares, are equal to fifty-eight dirhems and twenty things. Reduce this to one square, by taking the moiety of all you have. It is then: fifty dirhems and a square, which are equal to twenty-nine dirhems and ten things. Then reduce this, by taking twenty-nine from fifty; there remains twenty-one and a square, equal to ten things. Halve the number of the roots, it is five; multiply this by itself, it is twenty-five; take from this the twenty-one which are connected with the square, the remainder is four. Extract the root, it is two. Subtract this from the moiety of the roots, namely, from five, there remains three. This is one of the portions; the other is seven. This question refers you to one of the six cases, namely " squares and numbers equal to roots."

(29)

Sixth Problem.

I have multiplied one-third of a root by one-fourth of a root, and the product is equal to the root and twenty-four dirhems.*

$$* \quad \frac{x}{3} \times \frac{x}{4} = x + 24$$

$$\frac{x^2}{12} = x + 24$$

$$x^2 = 12x + 288$$

$$x = 6 + \sqrt{36 + 288} = 6 + 18 = 24$$

Computation : Call the root thing; then one-third of thing is multiplied by one-fourth of thing; this is the moiety of one-sixth of the square, and is equal to thing and twenty-four dirhems. Multiply this moiety of one-sixth of the square by twelve, in order to make your square a whole one, and multiply also the thing by twelve, which yields twelve things; and also four-and-twenty by twelve : the product of the whole will be two hundred and eighty-eight dirhems and twelve roots, which are equal to one square. The moiety of the roots is six. Multiply this by itself, and add it to two hundred and eighty-eight, it will be three hundred and twenty-four. Extract the root from this, it is eighteen; add this to the moiety of the roots, which was six ; the sum is twenty-four, and this is the square sought for. This question refers you to one of the six cases, namely, "roots and numbers equal to squares."

VARIOUS QUESTIONS.

IF a person puts such a question to you as : "I have (30) divided ten into two parts, and multiplying one of these by the other, the result was twenty-one;"* then

---

* $(10 - x)x = 21$
$10x - x^2 = 21$
which is to be reduced to
$x^2 + 21 = 10x$
$x = 5 \pm \sqrt{25 - 21} = 5 \pm 2$

you know that one of the two parts is thing, and the other ten minus thing. Multiply, therefore, thing by ten minus thing; then you have ten things minus a square, which is equal to twenty-one. Separate the square from the ten things, and add it to the twenty-one. Then you have ten things, which are equal to twenty-one dirhems and a square. Take away the moiety of the roots, and multiply the remaining five by itself; it is twenty-five. Subtract from this the twenty-one which are connected with the square; the remainder is four. Extract its root, it is two. Subtract this from the moiety of the roots, namely, five; there remain three, which is one of the two parts. Or, if you please, you may add the root of four to the moiety of the roots; the sum is seven, which is likewise one of the parts. This is one of the problems which may be resolved by addition and subtraction.

If the question be : "I have divided ten into two parts, and having multiplied each part by itself, I have subtracted the smaller from the greater, and the remainder was forty;"* then the computation is—you multiply ten (31) minus thing by itself, it is a hundred plus one square minus twenty things; and you also multiply thing by

---

$$* \quad (10-x)^2 - x^2 = 40$$
$$100 - 20x = 40$$
$$100 = 20x + 40$$
$$60 = 20x$$
$$3 = x$$

thing, it is one square. Subtract this from a hundred and a square minus twenty things, and you have a hundred, minus twenty things, equal to forty dirhems. Separate now the twenty things from a hundred, and add them to the forty; then you have a hundred, equal to twenty things and forty dirhems. Subtract now forty from a hundred; there remains sixty dirhems, equal to twenty things: therefore one thing is equal to three, which is one of the two parts.

If the question be: " I have divided ten into two parts, and having multiplied each part by itself, I have put them together, and have added to them the difference of the two parts previously to their multiplication, and the amount of all this is fifty-four;"* then the computation is this: You multiply ten minus thing by itself; it is a hundred and a square minus twenty things. Then multiply also the other thing of the ten by itself; it is one square. Add this together, it will be a hundred plus two squares minus twenty things. It was stated that the difference of the two parts before multiplication should be added to them. You say, therefore, the difference between them is ten minus two things.

---

* $(10-x)^2 + x^2 + (10-x) - x = 54$

$100 - 20x + 2x^2 + 10 - 2x = 54$

$100 - 22x + 2x^2 = 54$

$55 - 11x + x^2 = 27$

$x^2 + 28 = 11x$

$x = \frac{11}{2} \pm \sqrt{\frac{121}{4} - 28} = \frac{11 \pm 3}{2} = 7 \text{ or } 4$

The result is a hundred and ten and two squares minus twenty-two things, which are equal to fifty-four dirhems. Having reduced and equalized this, you may say, a hundred and ten dirhems and two squares are equal to fifty-four dirhems and twenty-two things. Reduce now the two squares to one square, by taking the moiety of all you have. Thus it becomes fifty-five dirhems and a square, equal to twenty-seven dirhems and eleven things. Subtract twenty-seven from fifty-five, there remain (32) twenty-eight dirhems and a square, equal to eleven things. Halve now the things, it will be five and a half; multiply this by itself, it is thirty and a quarter. Subtract from it the twenty-eight which are combined with the square, the remainder is two and a fourth. Extract its root, it is one and a half. Subtract this from the moiety of the roots, there remain four, which is one of the two parts.

If one say, "I have divided ten into two parts; and have divided the first by the second, and the second by the first, and the sum of the quotient is two dirhems and one-sixth;"* then the computation is this: If you multiply each part by itself, and add the products together, then their sum is equal to one of the parts

---

* $\dfrac{10-x}{x} + \dfrac{x}{10-x} = 2\frac{1}{6}$

$100 + 2x^2 - 20x = x(10-x) \times 2\frac{1}{6} = 21\frac{2}{3}x - 2\frac{1}{6}x^2$

$100 + 4\frac{1}{6}x^2 = 41\frac{2}{3}x$

$24 + x^2 = 10x$

$x = 5 \pm \sqrt{25-24} = 5 \pm 1 = 4 \text{ or } 6$

multiplied by the other, and again by the quotient which is two and one-sixth. Multiply, therefore, ten less thing by itself; it is a hundred and a square less ten things. Multiply thing by thing; it is one square. Add this together; the sum is a hundred plus two squares less twenty things, which is equal to thing multiplied by ten less thing; that is, to ten things less a square, multiplied by the sum of the quotients arising from the division of the two parts, namely, two and one-sixth. We have, therefore, twenty-one things and two-thirds of thing less two squares and one-sixth, equal to a hundred plus two squares less twenty things. Reduce this by adding the two squares and one-sixth to a hundred plus two squares less twenty things, and add the twenty negative things from the hundred plus the two squares to the twenty-one things and two-thirds of thing. Then you have a hundred plus four squares (33) and one-sixth of a square, equal to forty-one things and two-thirds of thing. Now reduce this to one square. You know that one square is obtained from four squares and one-sixth, by taking a fifth and one-fifth of a fifth.* Take, therefore, the fifth and one-fifth of a fifth of all that you have. Then it is twenty-four and a square, equal to ten roots; because ten is one-fifth and one-fifth of the fifth of the forty-one things and two-thirds of a thing. Now halve the roots; it gives five. Multiply this

---

* $4\frac{1}{6} = \frac{25}{6}$, and $\frac{6}{25} = \frac{1}{5} + \frac{1}{5} \times \frac{1}{5}$

by itself; it is five-and-twenty. Subtract from this the twenty-four, which are connected with the square; the remainder is one. Extract its root; it is one. Subtract this from the moiety of the roots, which is five. There remains four, which is one of the two parts.

Observe that, in every case, where any two quantities whatsoever are divided, the first by the second and the second by the first, if you multiply the quotient of the one division by that of the other, the product is always one.*

If some one say: "You divide ten into two parts; multiply one of the two parts by five, and divide it by the other: then take the moiety of the quotient, and add this to the product of the one part, multiplied by five; the sum is fifty dirhems;"† then the computation is this: Take thing, and multiply it by five. This is now to be divided by the remainder of the ten, that is, by ten less thing; and of the quotient the moiety is to be taken.

(34)   You know that if you divide five things by ten less thing, and take the moiety of the quotient, the result is

---

$$* \quad \frac{a}{b} \times \frac{b}{a} = 1$$

$$\dagger \quad \frac{5x}{2(10-x)} + 5x = 50$$

$$\frac{x}{2(10-x)} + x = 10$$

$$x^2 + 100 = 20\tfrac{1}{2}x$$

$$x = \tfrac{41}{4} - \tfrac{9}{4} = 8$$

the same as if you divide the moiety of five things by ten less thing. Take, therefore, the moiety of five things; it is two things and a half: and this you require to divide by ten less thing. Now these two things and a half, divided by ten less thing, give a quotient which is equal to fifty less five things: for the question states: add this (the quotient) to the one part multiplied by five, the sum will be fifty. You have already observed, that if the quotient, or the result of the division, be multiplied by the divisor, the dividend, or capital to be divided, is restored. Now, your capital, in the present instance, is two things and a half. Multiply, therefore, ten less thing by fifty less five things. Then you have five hundred dirhems and five squares less a hundred things, which are equal to two things and a half. Reduce this to one square. Then it becomes a hundred dirhems and a square less twenty things, equal to the moiety of thing. Separate now the twenty things from the hundred dirhems and square, and add them to the half thing. Then you have a hundred dirhems and a square, equal to twenty things and a half. Now halve the things, multiply the moiety by itself, subtract from this the hundred, extract the root of the remainder, and subtract this from the moiety of the roots, which is ten and one-fourth: the remainder is eight; and this is one of the portions.

If some one say: "You divide ten into two parts: multiply the one by itself; it will be equal to the other

taken eighty-one times."* Computation: You say, ten
less thing, multiplied by itself, is a hundred plus a
(35) square less twenty things, and this is equal to eighty-
one things. Separate the twenty things from a hundred
and a square, and add them to eighty-one. It will
then be a hundred plus a square, which is equal to a
hundred and one roots. Halve the roots; the moiety is
fifty and a half. Multiply this by itself, it is two thousand five hundred and fifty and a quarter. Subtract
from this one hundred; the remainder is two thousand
four hundred and fifty and a quarter. Extract the root
from this; it is forty-nine and a half. Subtract this
from the moiety of the roots, which is fifty and a half.
There remains one, and this is one of the two parts.

If some one say : " I have purchased two measures of
wheat or barley, each of them at a certain price. I
afterwards added the expences, and the sum was equal
to the difference of the two prices, added to the difference of the measures."†

---

* $(10-x)^2 = 81x$

$100 - 20x + x^2 = 81x$

$x^2 + 100 = 101x$

$x = \frac{101}{2} - \sqrt{\frac{101^2}{4} - 100} = 50\frac{1}{2} - 49\frac{1}{2} = 1$

† The purchaser does not make a clear enunciation of the
terms of his bargain. He intends to say, " I bought m
bushels of wheat, and n bushels of barley, and the wheat was
r times dearer than the barley. The sum I expended was
equal to the difference in the quantities, added to the difference in the prices of the grain."

Computation : Take what numbers you please, for it is indifferent; for instance, four and six. Then you say : I have bought each measure of the four for thing; and accordingly you multiply four by thing, which gives four things; and I have bought the six, each for the moiety of thing, for which I have bought the four; or, if you please, for one-third, or one-fourth, or for any other quota of that price, for it is indifferent. Suppose that you have bought the six measures for the moiety of thing, then you multiply the moiety of thing by six; this gives three things. Add them to the four things; the sum is seven things, which must be equal to the difference of the two quantities, which is two measures, plus the difference of the two prices, which is a moiety of thing. You have, therefore, seven things, equal to two and a moiety of thing. Remove, now, this moiety of thing, by subtracting it from the seven things. There remain six things and a half, equal to two dir- (36) hems: consequently, one thing is equal to four-thir- teenths of a dirhem. The six measures were bought, each at one-half of thing; that is, at two-thirteenths of a dirhem. Accordingly, the expenses amount to eight- and-twenty thirteenths of a dirhem, and this sum is equal to the difference of the two quantities; namely,

---

If x is the price of the barley, rx is the price of the wheat; whence, $mrx + nx = (m - n) + (rx - x)$; $\therefore x = \dfrac{m-n}{mr+n+r-1}$ and the sum expended is $\dfrac{(mr+n) \times (m-n)}{mr+n+r-1}$.

H

the two measures, the arithmetical equivalent for which is six-and-twenty thirteenths, added to the difference of the two prices, which is two-thirteenths: both differences together being likewise equal to twenty-eight parts.

If he say: "There are two numbers,* the difference of which is two dirhems. I have divided the smaller by the larger, and the quotient was just half a dirhem."† Suppose one of the two numbers* to be thing, and the other to be thing plus two dirhems. By the division of thing by thing plus two dirhems, half a dirhem appears as quotient. You have already observed, that by multiplying the quotient by the divisor, the capital which you divided is restored. This capital, in the present case, is thing. Multiply, therefore, thing and two dirhems by half a dirhem, which is the quotient; the product is half one thing plus one dirhem; this is equal to thing. Remove, now, half a thing on account

---

* In the original, " squares." The word square is used in the text to signify either, 1st, a square, properly so called, fractional or integral; 2d, a rational integer, not being a square number; 3d, a rational fraction, not being a square; 4th, a quadratic surd, fractional or integral.

$$\dagger \ \frac{x}{x+2} = \tfrac{1}{2}$$

$$x = \frac{x+2}{2} = \frac{x}{2} + 1$$

$$\frac{x}{2} = 1 \text{ and } x + 2 = 4$$

of the other half thing; there remains one dirhem, equal to half a thing. Double it, then you have one thing, equal to two dirhems. Consequently, the other number* is four.

If some one say: "I have divided ten into two parts; I have multiplied the one by ten and the other by itself, and the products were the same;"† then the computation is this: You multiply thing by ten; it is ten things. Then multiply ten less thing by itself; it is a hundred (37) and a square less twenty things, which is equal to ten things. Reduce this according to the rules, which I have above explained to you.

In like manner, if he say: " I have divided ten into two parts; I have multiplied one of the two by the other, and have then divided the product by the difference of the two parts before their multiplication, and the result of this division is five and one-fourth;"‡ the computation will be this: You subtract thing from ten; there remain ten less thing. Multiply the one by the other, it is ten things less a square. This is the product of the multiplication of one of the two parts by the other. At

---

* " Square " in the original.

† $10x = (10-x)^2 = 100 - 20x + x^2$

$x = 15 - \sqrt{225 - 100} = 15 - \sqrt{125}$

‡ $\dfrac{x(10-x)}{10-2x} = 5\frac{1}{4}$

$10x - x^2 = 52\frac{1}{2} - 10\frac{1}{2}x$

$20\frac{1}{2}x = x^2 + 52\frac{1}{2}$

$x = 10\frac{1}{4} - 7\frac{1}{4} = 3$

present you divide this by the difference between the
two parts, which is ten less two things. The quotient
of this division is, according to the statement, five and
a fourth. If, therefore, you muliply five and one-fourth
by ten less two things, the product must be equal to the
above amount, obtained by multiplication, namely, ten
things less one square. Multiply now five and one-
fourth by ten less two squares. The result is fifty-two
dirhems and a half less ten roots and a half, which is
equal to ten roots less a square. Separate now the ten
roots and a half from the fifty-two dirhems, and add
them to the ten roots less a square; at the same time
separate this square from them, and add it to the
fifty-two dirhems and a half. Thus you find twenty
roots and a half, equal to fifty-two dirhems and a half
and one square. Now continue reducing it, conform-
ably to the rules explained at the commencement of
this book.

(38)   If the question be: "There is a square, two-thirds of
one-fifth of which are equal to one-seventh of its root;"
then the square is equal to one root and half a seventh
of a root; and the root consists of fourteen-fifteenths
of the square.* The computation is this: You

$$* \ \tfrac{2}{3} \times \tfrac{1}{5}x^2 = \tfrac{x}{7}$$
$$x^2 = 7\tfrac{1}{2} \times \tfrac{x}{7} = 1\tfrac{1}{14}x$$
$$x = 1\tfrac{1}{14}$$
$$x^2 = 1\tfrac{29}{196}$$
$$\tfrac{2}{15}x^2 = \tfrac{30}{196} \times \tfrac{x}{7}$$

multiply two-thirds of one-fifth of the square by seven and a half, in order that the square may be completed. Multiply that which you have already, namely, one-seventh of its root, by the same. The result will be, that the square is equal to one root and half a seventh of the root; and the root of the square is one and a half seventh; and the square is one and twenty-nine one hundred and ninety-sixths of a dirhem. Two-thirds of the fifth of this are thirty parts of the hundred and ninety-six parts. One-seventh of its root is likewise thirty parts of a hundred and ninety-six.

If the instance be : " Three-fourths of the fifth of a square are equal to four-fifths of its root,"* then the computation is this : You add one-fifth to the four-fifths, in order to complete the root. This is then equal to three and three-fourths of twenty parts, that is, to fifteen eightieths of the square. Divide now eighty by fifteen ; the quotient is five and one-third. This is the root of the square, and the square is twenty-eight and four-ninths.

If some one say : " What is the amount of a square-root,† which, when multiplied by four times itself,

---

$$* \ \tfrac{3}{4} \times \tfrac{1}{5}x^2 = \tfrac{4}{5}x$$

$$\tfrac{3\frac{3}{4}}{20}x, \text{ or } \tfrac{15}{80}x, \text{ or } \tfrac{3}{16}x = 1$$

$$\therefore x = \tfrac{16}{3} = 5\tfrac{1}{3}.$$

† " Square " in the original.

amounts to twenty?*" the answer is this : If you multiply it by itself it will be five : it is therefore the root of five.

If somebody ask you for the amount of a square-root,† which when multiplied by its third amounts to ten,‡ the solution is, that when multiplied by itself it will amount to thirty ; and it is consequently the root of thirty.

(39)   If the question be: "To find a quantity†, which when multiplied by four times itself, gives one-third of the first quantity as product,"§ the solution is, that if you multiply it by twelve times itself, the quantity itself must re-appear : it is the moiety of one moiety of one-third.

If the question be: "A square, which when multiplied by its root gives three times the original square as product,"‖ then the solution is : that if you multiply the root by one-third of the square, the original square is

---

* $4x^2 = 20$
   $x = \sqrt{5}$

† " Square " in the original.

‡ $x \times \frac{x}{3} = 10$
   $x^2 = 30$
   $x = \sqrt{30}$

§ $x \times 4x = \frac{x}{3}$
   $x = \frac{1}{12}$

‖ $x^2 \times x = 3x^2$
   $x = 3$

restored; its root must consequently be three, and the square itself nine.

If the instance be: " To find a square, four roots of which, multiplied by three roots, restore the square with a surplus of forty-four dirhems,"* then the solution is: that you multiply four roots by three roots, which gives twelve squares, equal to a square and forty-four dirhems. Remove now one square of the twelve on account of the one square connected with the forty-four dirhems. There remain eleven squares, equal to forty-four dirhems. Make the division, the result will be four, and this is the square.

If the instance be: " A square, four of the roots of which multiplied by five of its roots produce twice the square, with a surplus of thirty-six dirhems;"† then the solution is: that you multiply four roots by five roots, which gives twenty squares, equal to two squares and thirty-six dirhems. Remove two squares from the twenty on account of the other two. The remainder is eighteen squares, equal to thirty-six dirhems. Divide now thirty-six dirhems by eighteen; the quotient is two, and this is the square.

$$* \quad 4x \times 3x = x^2 + 44$$
$$11x^2 = 44$$
$$x^2 = 4$$
$$x = 2$$
$$\dagger \quad 4x \times 5x = 2x^2 + 36$$
$$18x^2 = 36$$
$$x^2 = 2$$

(40)   In the same manner, if the question be: " A square, multiply its root by four of its roots, and the product will be three times the square, with a surplus of fifty dirhems."† Computation: You multiply the root by four roots, it is four squares, which are equal to three squares and fifty dirhems. Remove three squares from the four; there remains one square, equal to fifty dirhems. One root of fifty, multiplied by four roots of the same, gives two hundred, which is equal to three times the square, and a residue of fifty dirhems.

If the instance be: " A square, which when added to twenty dirhems, is equal to twelve of its roots,"† then the solution is this: You say, one square and twenty dirhems are equal to twelve roots. Halve the roots and multiply them by themselves; this gives thirty-six. Subtract from this the twenty dirhems, extract the root from the remainder, and subtract it from the moiety of the roots, which is six. The remainder is the root of the square: it is two dirhems, and the square is four.

If the instance be: "To find a square, of which if one-third be added to three dirhems, and the sum be subtracted from the square, the remainder multiplied by

---

$$* \ 4x^2 = 3x^2 + 50$$
$$x^2 = 50$$
$$† \ x^2 + 20 = 12x$$
$$x = 6 \pm \sqrt{36 - 20} = 6 \pm 4 = 10 \text{ or } 2$$

itself restores the square;"* then the computation
is this: If you subtract one-third and three dirhems
from the square, there remain two-thirds of it less three
dirhems. This is the root. Multiply therefore two-thirds
of thing less three dirhems by itself. You say two-
thirds by two-thirds is four ninths of a square; and less
two-thirds by three dirhems is two roots: and again,
two-thirds by three dirhems is two roots; and less three
dirhems by less three dirhems is nine dirhems. You (41)
have, therefore, four-ninths of a square and nine dirhems
less four roots, which are equal to one root. Add the
four roots to the one root, then you have five roots,
which are equal to four-ninths of a square and nine
dirhems. Complete now your square; that is, multiply
the four-ninths of a square by two and a fourth, which
gives one square; multiply likewise the nine dirhems
by two and a quarter; this gives twenty and a quarter;
finally, multiply the five roots by two and a quarter;
this gives eleven roots and a quarter. You have, therefore, a square and twenty dirhems and a quarter, equal
to eleven roots and a quarter. Reduce this according to
what I taught you about halving the roots.

---

* $\left[x-\left(\frac{x}{3}+3\right)\right]^2=x$

or $\left[\frac{2x}{3}-3\right]^2=x$

$\frac{4x^2}{9}+9=5x$

$x^2+20\frac{1}{4}=11\frac{1}{4}x$

$x=9,$ or $2\frac{1}{4}$

If the instance be: "To find a number,* one-third of which, when multiplied by one-fourth of it, restores the *number,"† then the computation is: You multiply one-third of thing by one-fourth of thing, this gives one-twelfth of a square, equal to thing, and the square is equal to twelve things, which is the root of one hundred and forty-four.

If the instance be: "A number,* one-third of which and one dirhem multiplied by one-fourth of it and two dirhems restore the number,* with a surplus of thirteen dirhems;"‡ then the computation is this: You multiply one-third of thing by one-fourth of thing, this gives half one-sixth of a square; and you multiply two dirhems by one-third of thing, this gives two-thirds of a root; and one dirhem by one-fourth of thing gives one-fourth of a root; and one dirhem by two dirhems gives two dirhems. This altogether is one-twelfth of a square and two dirhems and eleven-
(42) twelfths of a thing, equal to thing and thirteen dir-

---

* " Square " in the original.

$$† \frac{x}{3} \times \frac{x}{4} = x$$
$$x^2 = 12x$$
$$x = 12$$

$$‡ \left(\frac{x}{3}+1\right)-\left(\frac{x}{4}+2\right)=x+13$$
$$\frac{x^2}{12}+1\tfrac{1}{12}x + 2 = x + 13$$
$$\frac{x^2}{12} = \frac{x}{12} + 11$$
$$x^2 = x + 132$$
$$x = \tfrac{1}{2} + \tfrac{23}{2} = 12$$

hems. Remove now two dirhems from thirteen, on account of the other two dirhems, the remainder is eleven dirhems. Remove then the eleven-twelfths of a root from the one (root on the opposite side), there remains one-twelfth of a root and eleven dirhems, equal to one-twelfth of a square. Complete the square: that is, multiply it by twelve, and do the same with all you have. The product is a square, which is equal to a hundred and thirty-two dirhems and one root. Reduce this, according to what I have taught you, it will be right.

If the instance be: "A dirhem and a half to be divided among one person and certain persons, so that the share of the one person be twice as many dirhems as there are other persons;"* then the Computation is this:† You say, the one person and some persons are one and thing: it is the same as if the question had been one dirhem and a half to be divided by one and thing, and the share of one person to be equal to two things. Multiply, therefore, two things by one and

---

* The enunciation in the original is faulty, and I have altered it to correspond with the computation. But in the computation, x, the number of persons, is fractional! I am unable to correct the passage satisfactorily.

$$\dagger \ \frac{1\frac{1}{2}}{1+x} = 2x$$
$$x2 + x = \frac{3}{4}$$
$$x = 1 - \frac{1}{2}$$
$$x = \frac{1}{2}$$

thing; it is two squares and two things, equal to one dirhem and a half. Reduce them to one square: that is, take the moiety of all you have. You say, therefore, one square and one thing are equal to three-fourths of a dirhem. Reduce this, according to what I have taught you in the beginning of this work.

If the instance be: "A number,[*] you remove one-third of it, and one-fourth of it, and four dirhems: then you multiply the remainder by itself, and the number,[*] is restored, with a surplus of twelve dirhems:"[†] then the computation is this: You take thing, and subtract from it one-third and one-fourth; there remain five-twelfths of thing. Subtract from this four dirhems: (43) the remainder is five-twelfths of thing less four dirhems. Multiply this by itself. Thus the five parts become five-and-twenty parts; and if you multiply twelve by itself, it is a hundred and forty-four. This makes, therefore, five and twenty hundred and forty-fourths of a square. Multiply then the four dirhems twice by the five-twelfths; this gives forty parts, every twelve of which make one root (forty-twelfths); finally, the four

---

* " Square" in the original.

$$† \ (x-\tfrac{1}{3}x-\tfrac{1}{4}x-4)^2=x+12$$
$$(\tfrac{5}{12}x-4)^2=x+12$$
$$\tfrac{25}{144}x^2+16-3\tfrac{1}{3}x=x+12$$
$$\tfrac{25}{144}x^2+4=4\tfrac{1}{3}x$$
$$x^2+23\tfrac{1}{25}=24\tfrac{24}{25}x$$
$$\sqrt{\left[\left(\tfrac{24\frac{24}{25}}{2}\right)^2-23\tfrac{1}{25}\right]}+\tfrac{24\frac{24}{25}}{2}-x$$
$$11\tfrac{13}{25}+12\tfrac{12}{25}=24=x$$

dirhems, multiplied by four dirhems, give sixteen dirhems to be added. The forty-twelfths are equal to three roots and one-third of a root, to be subtracted. The whole product is, therefore, twenty-five-hundred-and-forty-fourths of a square and sixteen dirhems less three roots and one-third of a root, equal to the original number,* which is thing and twelve dirhems. Reduce this, by adding the three roots and one-third to the thing and twelve dirhems. Thus you have four roots and one-third of a root and twelve dirhems. Go on balancing, and subtract the twelve (dirhems) from sixteen; there remain four dirhems and five-and-twenty-hundred-and-forty-fourths of a square, equal to four roots and one-third. Now it is necessary to complete the square. This you can accomplish by multiplying all you have by five and nineteen twenty-fifths. Multiply, therefore, the twenty-five-one-hundred-and-forty-fourths of a square by five and nineteen twenty-fifths. This gives a square. Then multiply the four (44) dirhems by five and nineteen twenty-fifths; this gives twenty-three dirhems and one twenty-fifth. Then multiply four roots and one-third by five and nineteen twenty-fifths; this gives twenty-four roots and twenty-four twenty-fifths of a root. Now halve the number of the roots: the moiety is twelve roots and twelve twenty-fifths of a root. Multiply this by itself. It is one hundred-and-fifty-five dirhems and four hundred-and-

* " Square " in the original.

sixty-nine six-hundred-and-twenty-fifths. Subtract from this the twenty-three dirhems and the one twenty-fifth connected with the square. The remainder is one-hundred-and-thirty-two and four-hundred-and-forty six-hundred-and-twenty-fifths. Take the root of this: it is eleven dirhems and thirteen twenty-fifths. Add this to the moiety of the roots, which was twelve dirhems and twelve twenty-fifths. The sum is twenty-four. It is the number* which you sought. When you subtract its third and its fourth and four dirhems, and multiply the remainder by itself, the number * is restored, with a surplus of twelve dirhems.

If the question be: " To find a square-root,* which, when multiplied by two-thirds of itself, amounts to (45) five;"† then the computation is this: You multiply one thing by two-thirds of thing; the product is two-thirds of square, equal to five. Complete it by adding its moiety to it, and add to five likewise its moiety. Thus you have a square, equal to seven and a half. Take its root; it is the thing which you required, and which, when multiplied by two-thirds of itself, is equal to five.

If the instance be: " Two numbers,‡ the difference

---

* " Square " in the original.

$$\dagger \quad x \times \tfrac{2}{3}x = 5$$
$$\tfrac{2}{3}x^2 = 5$$
$$x^2 = 7\tfrac{1}{2}$$
$$x = \sqrt{7\tfrac{1}{2}}$$

‡ " Squares " in the original.

of which is two dirhems; you divide the small one by the great one, and the quotient is equal to half a dirhem;* then the computation is this: Multiply thing and two dirhems by the quotient, that is a half. The product is half a thing and one dirhem, equal to thing. Remove now half a dirhem on account of the half dirhem on the other side. The remainder is one dirhem, equal to half a thing. Double it: then you have thing, equal to two dirhems. This is one of the two numbers,† and the other is four.

Instance: " You divide one dirhem amongst a certain number of men, which number is thing. Now you add one man more to them, and divide again one dirhem amongst them; the quota of each is then one-sixth of a dirhem less than at the first time."‡ Computation: You multiply the first number of men, which is thing, by the difference of the share for each of the other number. Then multiply the product by the first and second number of men, and divide the product by the

---

$$* \quad \frac{x}{x+2} = \frac{1}{2}$$
$$\tfrac{1}{2}x + 1 = x$$
$$\tfrac{1}{2}x = 1$$
$$x = 2, \ x + 2 = 4$$

† " Squares " in the original.

$$‡ \quad \frac{1}{x} - \frac{1}{x+1} = \frac{1}{6}$$
$$1 = \frac{x(x+1)}{6}$$
$$x^2 + x = 6$$
$$\sqrt{[\tfrac{1}{2}]^2 + 6} - \tfrac{1}{2} = x = 2$$

difference of these two numbers. Thus you obtain the sum which shall be divided. Multiply, therefore, the first number of men, which is thing, by the one-sixth, which is the difference of the shares; this gives one-sixth of root. Then multiply this by the original number of the men, and that of the additional one, that is to say, by thing plus one. The product is one-sixth of square and one-sixth of root divided by one

(46) dirhem, and this is equal to one dirhem. Complete the square which you have through multiplying it by six. Then you have a square and a root equal to six dirhems. Halve the root and multiply the moiety by itself, it is one-fourth. Add this to the six; take the root of the sum and subtract from it the moiety of the root, which you have multiplied by itself, namely, a half. The remainder is the first number of men; which in this instance is two.

If the instance be : " To find a square-root,* which when multiplied by two-thirds of itself amounts to five:"† then the computation is this: If you multiply it by itself, it gives seven and a half. Say, therefore,

---

* " Square " in the original.

$$\dagger \; \tfrac{2}{3}x^2 = 5$$
$$x^2 = 7\tfrac{1}{2}$$
$$x = \sqrt{7\tfrac{1}{2}}$$
$$\sqrt{7\tfrac{1}{2}} \times \tfrac{2}{3}\sqrt{7\tfrac{1}{2}} = 5$$
$$\sqrt{\tfrac{4}{9} \times 7\tfrac{1}{2}} = \sqrt{3\tfrac{1}{3}} = \tfrac{2}{3}\sqrt{7\tfrac{1}{2}}$$
$$\sqrt{3\tfrac{1}{3} \times 7\tfrac{1}{2}} = \sqrt{25} = 5$$

it is the root of seven and a half multiplied by two-thirds of the root of seven and a half. Multiply then two-thirds by two-thirds, it is four-ninths; and four-ninths multiplied by seven and a half is three and a third. The root of three and a third is two-thirds of the root of seven and a half. Multiply three and a third by seven and a half; the product is twenty-five, and its root is five.

If the instance be : "A square multiplied by three of its roots is equal to five times the original square;"* then this is the same as if it had been said, a square, which when multiplied by its root, is equal to the first square and two-thirds of it. Then the root of the square is one and two-thirds, and the square is two dirhems and seven-ninths.

If the instance be : " Remove one-third from a square, then multiply the remainder by three roots of the first square, and the first square will be restored."†
Computation : If you multiply the first square, before (47) removing two-thirds from it, by three roots of the same, then it is one square and a half; for according to the statement two-thirds of it multiplied by three

---

$$* \ x^2 \times 3x = 5x^2$$
$$x^2 \times x = 1\tfrac{2}{3}x^2$$
$$x = 1\tfrac{2}{3}$$
$$x^2 = 2\tfrac{7}{9}$$
$$\dagger \ (x^2 - \tfrac{1}{3}x^2) \times 3x = x^2 \ \therefore \ \tfrac{2}{3}x^2 \times 3x = x^2$$
$$x^2 \times 3x = 1\tfrac{1}{2}x^2$$
$$x = \tfrac{1}{2} \ \therefore \ x^2 = \tfrac{1}{4}$$

K

roots give one square; and, consequently, the whole of it multiplied by three roots of it gives one square and a half. This entire square, when multiplied by one root, gives half a square; the root of the square must therefore be a half, the square one-fourth, two-thirds of the square one-sixth, and three roots of the square one and a half. If you multiply one-sixth by one and a half, the product is one-fourth, which is the square.

Instance: "A square; you subtract four roots of the same, then take one-third of the remainder; this is equal to the four roots." The square is two hundred and fifty-six.* Computation: You know that one-third of the remainder is equal to four roots; consequently, the whole remainder must be twelve roots; add to this the four roots; the sum is sixteen, which is the root of the square.

Instance: "A square; you remove one root from it; and if you add to this root a root of the remainder, the sum is two dirhems."† Then, this is the root of a

---

$$* \ \frac{x^2 - 4x}{3} = 4x$$
$$x^2 - 4x = 12x$$
$$x^2 = 16x$$
$$x = 16 \ \therefore \ x^2 = 256$$
$$† \ \sqrt{x^2 - x} + x = 2$$
$$\sqrt{x^2 - x} = 2 - x$$
$$x^2 - x = 4 + x^2 - 4x$$
$$x^2 + 3x = 4 + x^2$$
$$3x = 4$$
$$x = 1\tfrac{1}{3}$$

square, which, when added to the root of the same square, less one root, is equal to two dirhems. Subtract from this one root of the square, and subtract also from the two dirhems one root of the square. Then two dirhems less one root multiplied by itself is four dirhems and one square less four roots, and this is equal to a square less one root. Reduce it, and you find a square and four dirhems, equal to a square and three roots. Remove square by square; there remain three roots, equal to four dirhems; consequently, one root is equal to one dirhem and one-third. This is the root of the square, and the square is one dirhem and seven-ninths of a dirhem. (48)

Instance: " Subtract three roots from a square, then multiply the remainder by itself, and the square is restored."* You know by this statement that the remainder must be a root likewise; and that the square consists of four such roots; consequently, it must be sixteen.

$$* \ (x^2 - 3x)^2 = x^2$$
$$x^2 - 3x = x$$
$$x^2 = 4x$$
$$x = 4$$

ON MERCANTILE TRANSACTIONS.

You know that all mercantile transactions of people, such as buying and selling, exchange and hire, comprehend always two notions and four numbers, which are stated by the enquirer; namely, measure and price, and quantity and sum. The number which expresses the measure is inversely proportionate to the number which expresses the sum, and the number of the price inversely proportionate to that of the quantity. Three of these four numbers are always known, one is unknown, and this is implied when the person inquiring says *how much?* and it is the object of the question. The computation in such instances is this, that you try the three given numbers; two of them must necessarily be inversely proportionate the one to the other. Then you multiply these two proportionate numbers by each other, and you divide the product by the third given number, the proportionate of which is unknown. The quotient of this division is the unknown number, which the inquirer asked for; and it is inversely proportionate to the divisor.*

Examples.—For the first case : If you are told, " ten
(49) for six, how much for four ?" then *ten* is the measure ;

---

* If a is given for b, and A for B, then $a : b :: A : B$ or $aB = bA \therefore a = \dfrac{bA}{B}$ and $b = \dfrac{aB}{A}$.

six is the price; the expression *how much* implies the unknown number of the quantity; and *four* is the number of the sum. The number of the measure, which is *ten*, is inversely proportionate to the number of the sum, namely, *four*. Multiply, therefore, ten by four, that is to say, the two known proportionate numbers by each other; the product is forty. Divide this by the other known number, which is that of the price, namely, six. The quotient is six and two-thirds; it is the unknown number, implied in the words of the question " *how much ?*" it is the quantity, and inversely proportionate to the six, which is the price.

For the second case : Suppose that some one ask this question : " ten for eight, what must be the sum for four ?" This is also sometimes expressed thus : " What must be the price of four of them ?" Ten is the number of the measure, and is inversely proportionate to the unknown number of the sum, which is involved in the expression *how much* of the statement. Eight is the number of the price, and this is inversely proportionate to the known number of the quantity, namely, four. Multiply now the two known proportionate numbers one by the other, that is to say, four by eight. The product is thirty-two. Divide this by the other known number, which is that of the measure, namely, ten. The quotient is three and one-fifth; this is the number of the sum, and inversely proportionate to the ten which was the divisor. In this manner all computations in matters of business may be solved.

If somebody says, "a workman receives a pay of ten
(50) dirhems per month ; how much must be his pay for six
days ?" Then you know that six days are one-fifth of
the month; and that his portion of the dirhems must
be proportionate to the portion of the month. You
calculate it by observing that one month, or thirty
days, is the measure, ten dirhems the price, six days
the quantity, and his portion the sum. Multiply the
price, that is, ten, by the quantity, which is proportionate to it, namely, six; the product is sixty. Divide
this by thirty, which is the known number of the measure. The quotient is two dirhems, and this is the sum.

This is the proceeding by which all transactions con-
cerning exchange or measures or weights are settled.

MENSURATION.

KNOW that the meaning of the expression "one by
one" is mensuration : one yard (in length) by one yard
(in breadth) being understood.

Every quadrangle of equal sides and angles, which
has one yard for every side, has also *one* for its area.
Has such a quadrangle two yards for its side, then the
area of the quadrangle is four times the area of a quadrangle, the side of which is one yard. The same takes
place with three by three, and so on, ascending or
descending: for instance, a half by a half, which gives

a quarter, or other fractions, always following the same rule. A quadrate, every side of which is half a yard, is (51) equal to one-fourth of the figure which has one yard for its side. In the same manner, one-third by one-third, or one-fourth by one-fourth, or one-fifth by one-fifth, or two-thirds by a half, or more or less than this, always according to the same rule.

One side of an equilateral quadrangular figure, taken once, is its root; or if the same be multiplied by two, then it is like two of its roots, whether it be small or great.

If you multiply the height of any equilateral triangle by the moiety of the basis upon which the line marking the height stands perpendicularly, the product gives the area of that triangle.

In every equilateral quadrangle, the product of one diameter multiplied by the moiety of the other will be equal to the area of it.

In any circle, the product of its diameter, multiplied by three and one-seventh, will be equal to the periphery. This is the rule generally followed in practical life, though it is not quite exact. The geometricians have two other methods. One of them is, that you multiply the diameter by itself; then by ten, and hereafter take the root of the product; the root will be the periphery. The other method is used by the astronomers among them: it is this, that you multiply the diameter by sixty-two thousand eight hundred and thirty-two and then divide the product by twenty thousand; the quotient is the periphery. Both methods come very nearly to the same effect.*

If you divide the periphery by three and one-seventh, the quotient is the diameter.

The area of any circle will be found by multiplying the moiety of the circumference by the moiety of the diameter; since, in every polygon of equal sides and (52) angles, such as triangles, quadrangles, pentagons, and so on, the area is found by multiplying the moiety of the circumference by the moiety of the diameter of the middle circle that may be drawn through it.

If you multiply the diameter of any circle by itself, and subtract from the product one-seventh and half one-seventh of the same, then the remainder is equal to the area of the circle. This comes very nearly to the same result with the method given above. †

Every part of a circle may be compared to a bow. It must be either exactly equal to half the circumference, or less or greater than it. This may be ascertained by the arrow of the bow. When this becomes equal to the moiety of the chord, then the arc is

---

* The three formulas are,

$$\text{1st, } 3\tfrac{1}{7}d = p \text{ i.e. } 3.1428\,d$$

$$\text{2d, } \sqrt{10d^2} = p \text{ i.e. } 3.16227\,d$$

$$\text{3d, } \frac{d \times 62832}{20000} = p \text{ i.e. } 3.1416\,d$$

† The area of a circle whose diameter is d is $\pi\frac{d^2}{4} = \frac{22}{7 \times 4}d^2 = \left(1 - \tfrac{1}{7} - \frac{1}{2 \times 7}\right)d^2.$

exactly the moiety of the circumference: is it shorter than the moiety of the chord, then the bow is less than half the circumference; is the arrow longer than half the chord, then the bow comprises more than half the circumference.

If you want to ascertain the circle to which it belongs, multiply the moiety of the chord by itself, divide it by the arrow, and add the quotient to the arrow, the sum is the diameter of the circle to which this bow belongs.

If you want to compute the area of the bow, multiply the moiety of the diameter of the circle by the moiety of the bow, and keep the product in mind. Then subtract the arrow of the bow from the moiety of the diameter of the circle, if the bow is smaller than half the circle; or if it is greater than half the circle, subtract half the diameter of the circle from the arrow of the bow. Multiply the remainder by the moiety of the chord of the bow, and subtract the product from that which you have kept in mind if the bow is smaller (53) than the moiety of the circle, or add it thereto if the bow is greater than half the circle. The sum after the addition, or the remainder after the subtraction, is the area of the bow.

The bulk of a quadrangular body will be found by multiplying the length by the breadth, and then by the height.

If it is of another shape than the quadrangular (for instance, circular or triangular), so, however, that a line representing its height may stand perpendicularly on its basis, and yet be parallel to the sides, you must calculate it by ascertaining at first the area of its basis. This, multiplied by the height, gives the bulk of the body.

Cones and pyramids, such as triangular or quadrangular ones, are computed by multiplying one-third of the area of the basis by the height.

Observe, that in every rectangular triangle the two short sides, each multiplied by itself and the products added together, equal the product of the long side multiplied by itself.

The proof of this is the following. We draw a quadrangle, with equal sides and angles A B C D. We divide the line A C into two moieties in the point H, from which we draw a parallel to the point R. Then we divide, also, the line A B into two moieties at the point T, and draw a parallel to the point G. Then the quadrate A B C D is divided into four quadrangles of equal sides and angles, and of equal area; namely, the squares A K, C K, B K, and D K. Now, we draw from (54) the point H to the point T a line which divides the quadrangle A K into two equal parts: thus there arise two triangles from the quadrangle, namely, the triangles A T H and H K T. We know that A T is the moiety of A B, and that A H is equal to it, being the moiety of A C; and the line T H joins them opposite the right angle. In the same manner we draw lines from T to R, and from R to G, and from G to H. Thus from all the squares eight equal triangles arise, four of which must, consequently, be equal to the moiety of the great quadrate A D. We know that the line AT multiplied by itself is like the area of two triangles, and A K gives the area of two triangles equal to them; the sum of them is therefore four triangles. But the line HT, multiplied by itself, gives likewise the area of four such triangles. We perceive, therefore, that the sum of AT multiplied by itself, added to A H multiplied by itself, is equal to TH multiplied by itself. This is the observation which we were desirous to elucidate. Here is the figure to it:

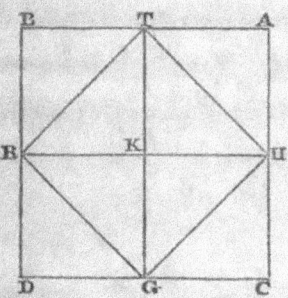

Quadrangles are of five kinds: firstly, with right (55) angles and equal sides; secondly, with right angles and unequal sides; thirdly, the rhombus, with equal sides and unequal angles; fourthly, the rhomboid, the length of which differs from its breadth, and the angles of which are unequal, only that the two long and the two short sides are respectively of equal length; fifthly, quadrangles with unequal sides and angles.

First kind.—The area of any quadrangle with equal sides and right angles, or with unequal sides and right angles, may be found by multiplying the length by the breadth. The product is the area. For instance: a quadrangular piece of ground, every side of which has five yards, has an area of five-and-twenty square yards. Here is its figure.

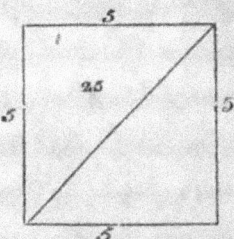

Second kind.—A quadrangular piece of ground, the two long sides of which are of eight yards each, while the breadth is six. You find the area by multiplying six by eight, which yields forty-eight yards. Here is (56) the figure to it:

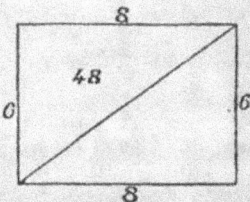

Third kind, the Rhombus.—Its sides are equal: let each of them be five, and let its diagonals be, the one eight and the other six yards. You may then compute the area, either from one of the diagonals, or from both. As you know them both, you multiply the one by the moiety of the other, the product is the area: that is to say, you multiply eight by three, or six by four; this yields twenty-four yards, which is the area.

If you know only one of the diagonals, then you are aware, that there are two triangles, two sides of each of which have every one five yards, while the third is the diagonal. Hereafter you can make the computation according to the rules for the triangles.* This is the figure:

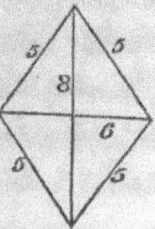

The fourth kind, or Rhomboid, is computed in the same way as the rhombus. Here is the figure to it:

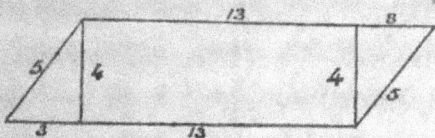

The other quadrangles are calculated by drawing a (57) diagonal, and computing them as triangles.

Triangles are of three kinds, acute-angular, obtuse-angular, or rectangular. The peculiarity of the rectangular triangle is, that if you multiply each of its two short sides by itself, and then add them together, their sum will be equal to the long side multiplied by itself. The character of the acute-angled triangle is

---

* If the two diagonals are d and d', and the side s, the area of the rhombus is $\frac{dd'}{2} = d \times \sqrt{s^2 - \frac{d^2}{4}}$.

this: if you multiply every one of its two short sides by itself, and add the products, their sum is more than the long side alone multiplied by itself. The definition of the obtuse-angled triangle is this: if you multiply its two short sides each by itself, and then add the products, their sum is less than the product of the long side multiplied by itself.

The rectangular triangle has two cathetes and an hypotenuse. It may be considered as the moiety of a quadrangle. You find its area by multiplying one of its cathetes by the moiety of the other. The product is the area.

Examples.—A rectangular triangle; one cathete being (58) six yards, the other eight, and the hypotenuse ten. You make the computation by multiplying six by four: this gives twenty-four, which is the area. Or if you prefer, you may also calculate it by the height, which rises perpendicularly from the longest side of it: for the two short sides may themselves be considered as two heights. If you prefer this, you multiply the height by the moiety of the basis. The product is the area. This is the figure:

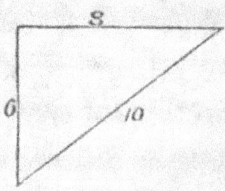

Second kind.— An equilateral triangle with acute angles, every side of which is ten yards long. Its area

may be ascertained by the line representing its height
and the point from which it rises.* Observe, that in
every isosceles triangle, a line to represent the height
drawn to the basis rises from the latter in a right
angle, and the point from which it proceeds is always
situated in the midst of the basis; if, on the contrary,
the two sides are not equal, then this point never lies
in the middle of the basis. In the case now before us
we perceive, that towards whatever side we may draw
the line which is to represent the height, it must
necessarily always fall in the middle of it, where the
length of the basis is five. Now the height will be
ascertained thus. You multiply five by itself; then
multiply one of the sides, that is ten, by itself, which
gives a hundred. Now you subtract from this the
product of five multiplied by itself, which is twenty-five. (59)
The remainder is seventy-five, the root of which is the
height. This is a line common to two rectangular triangles. If you want to find the area, multiply the
root of seventy-five by the moiety of the basis, which is
five. This you perform by multiplying at first five by
itself; then you may say, that the root of seventy-five is
to be multiplied by the root of twenty-five. Multiply
seventy-five by twenty-five. The product is one thousand eight hundred and seventy-five; take its root, it is

---

* The height of the equilateral triangle whose side is 10,
is $\sqrt{10^2 - 5^2} = \sqrt{75}$, and the area of the triangle is
$5\sqrt{75} = 25\sqrt{3}$

the area: it is forty-three and a little.*   Here is the figure:

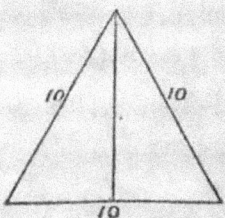

There are also acute-angled triangles, with different sides. Their area will be found by means of the line representing the height and the point from which it proceeds. Take, for instance, a triangle, one side of which is fifteen yards, another fourteen, and the third thirteen yards. In order to find the point from which the line marking the height does arise, you may take for the basis any side you choose; e. g. that which is fourteen yards long. The point from which the line (60) representing the height does arise, lies in this basis at an unknown distance from either of the two other sides. Let us try to find its unknown distance from the side which is thirteen yards long. Multiply this distance by itself; it becomes an [unknown] square. Subtract this from thirteen multiplied by itself; that is, one hundred and sixty-nine. The remainder is one hundred and sixty-nine less a square. The root from this is the height. The remainder of the basis is fourteen less thing. We multiply this by itself; it becomes one hundred and ninety-six, and a square less twenty-

---

* The root is 43. 3 +

eight things. We subtract this from fifteen multiplied by itself; the remainder is twenty-nine dirhems and twenty-eight things less one square. The root of this is the height. As, therefore, the root of this is the height, and the root of one hundred and sixty-nine less square is the height likewise, we know that they both are the same.* Reduce them, by removing square against square, since both are negatives. There remain twenty-nine [dirhems] plus twenty-eight things, which are equal to one hundred and sixty-nine. Subtract now twenty-nine from one hundred and sixty-nine. The remainder is one hundred and forty, equal to twenty-eight things. One thing is, consequently, five. This is the distance of the said point from the side of thirteen yards. The complement of the basis towards the other side is nine. Now in order to find the height, you multiply five by itself, and subtract it from the contiguous side, which is thirteen, multiplied by itself. The remainder is one hundred and forty-four. Its root is the height. It is twelve. The height forms always two (61) right angles with the basis, and it is called the *column*, on account of its standing perpendicularly. Multiply the height into half the basis, which is seven. The

---

$$* \sqrt{169} - x^2 = 29 + 28x - x^2$$
$$163 = 29 + 28x$$
$$140 = 28x$$
$$5 = x$$

M

product is eighty-four, which is the area. Here is the figure :

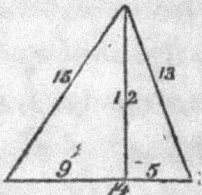

The third species is that of the obtuse-angled triangle with one obtuse angle and sides of different length. For instance, one side being six, another five, and the third nine. The area of such a triangle will be found by means of the height and of the point from which a line representing the same arises. This point can, within such a triangle, lie only in its longest side. Take therefore this as the basis : for if you choose to take one of the short sides as the basis, then this point would fall beyond the triangle. You may find the distance of this point, and the height, in the same manner, which I have shown in the acute-angled triangle; the whole computation is the same. Here is the figure :

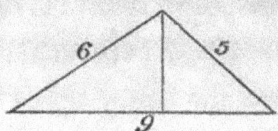

We have above treated at length of the circles, of their qualities and their computation. The following (62) is an example : If a circle has seven for its diameter, then it has twenty-two for its circumference. Its area you find in the following manner : Multiply the moiety of the diameter, which is three and a half, by the moiety of the circumference, which is eleven. The product is thirty-eight and a half, which is the area. Or you may also multiply the diameter, which is seven, by itself: this is forty-nine; subtracting herefrom one-seventh and half one-seventh, which is ten and a half, there remain thirty-eight and a half, which is the area. Here is the figure:

If some one inquires about the bulk of a pyramidal pillar, its base being four yards by four yards, its height ten yards, and the dimensions at its upper extremity two yards by two yards; then we know already that every pyramid is decreasing towards its top, and that one-third of the area of its basis, multiplied by the height, gives its bulk. The present pyramid has no top. We must therefore seek to ascertain what is wanting in its height to complete the top. We observe, that the proportion of the entire height to the ten, which we have now before us, is equal to the proportion of four to two. Now as two is the moiety of four, ten must likewise be the moiety of the entire height, and the whole height of the pillar must be twenty yards. At present we take one-third of the area of the basis, that is, five and one-third, and multiply it by the length, which is twenty. The product is one hundred (63)

and six yards and two-thirds. Herefrom we must then subtract the piece, which we have added in order to complete the pyramid. This we perform by multiplying one and one-third, which is one-third of the product of two by two, by ten: this gives thirteen and a third. This is the piece which we have added in order to complete the pyramid. Subtracting this from one hundred and six yards and two-thirds, there remain ninety-three yards and one-third: and this is the bulk of the mutilated pyramid. This is the figure:

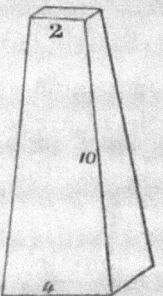

If the pillar has a circular basis, subtract one-seventh and half a seventh from the product of the diameter multiplied by itself, the remainder is the basis.

If some one says: "There is a triangular piece of land, two of its sides having ten yards each, and the basis twelve; what must be the length of one side of a quadrate situated within such a triangle?" the solution is this. At first you ascertain the height of the triangle, by multiplying the moiety of the basis, (which is six) by itself, and subtracting the product, which is thirty-six, from one of the two short sides multiplied by itself, which is one hundred; the remainder is sixty-four: take the root from this; it is eight. This (64) is the height of the triangle. Its area is, therefore, forty-eight yards: such being the product of the height multiplied by the moiety of the basis, which is six. Now we assume that one side of the quadrate inquired for is thing. We multiply it by itself; thus it becomes a square, which we keep in mind. We know that there must remain two triangles on the two sides of the quadrate, and one above it. The two triangles on both sides of it are equal to each other: both having the same height and being rectangular. You find their area by multiplying thing by six less half a thing, which gives six things less half a square. This is the area of both the triangles on the two sides of the quadrate together. The area of the upper triangle will be found by multiplying eight less thing, which is the height, by half one thing. The product is four things less half a square. This altogether is equal to the area of the quadrate plus that of the three triangles: or, ten things are equal to forty-eight, which is the area of the great triangle. One thing from this is four yards and four-fifths of a yard; and this is the length of any side of the quadrate. Here is the figure:

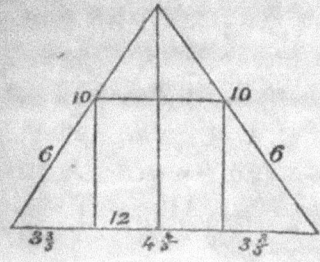

ON LEGACIES.

———

On Capital, and Money lent.

(65)   " A man dies, leaving two sons behind him, and bequeathing one-third of his capital to a stranger. He leaves ten dirhems of property and a claim of ten dirhems upon one of the sons."

Computation: You call the sum which is taken out of the debt thing. Add this to the capital, which is ten dirhems. The sum is ten and thing. Subtract one-third of this, since he has bequeathed one-third of his property, that is, three dirhems and one-third of thing. The remainder is six dirhems and two-thirds of thing. Divide this between the two sons. The portion of each of them is three dirhems and one-third plus one-third of thing. This is equal to the thing which was sought for.* Reduce it, by removing one-third from

———

* If a father dies, leaving n sons, one of whom owes the father a sum exceeding an nth part of the residue of the father's estate, after paying legacies, then such son retains the whole sum which he owes the father: part, as a set-off against his share of the residue, the surplus as a gift from the father.

In the present example, let each son's share of the residue be equal to x.

$\frac{2}{3} [10 + x] = 2x$ $\therefore$ $1 + x - 3x$ $\therefore$ $10 = 2x$ $\therefore$ $x = 5$. The stranger receives 5; and the son, who is not indebted to the father, receives 5.

thing, on account of the other third of thing. There remain two-thirds of thing, equal to three dirhems and one-third. It is then only required that you complete the thing, by adding to it as much as one half of the same; accordingly, you add to three and one-third as much as one-half of them: This gives five dirhems, which is the thing that is taken out of the debts.

If he leaves two sons and ten dirhems of capital and a demand of ten dirhems against one of the sons, and bequeaths one-fifth of his property and one dirhem to a stranger, the computation is this: Call the sum which is taken out of the debt, thing. Add this to the property; the sum is thing and ten dirhems. Subtract one-fifth of this, since he has bequeathed one-fifth of (66) his capital, that is, two dirhems and one-fifth of thing; the remainder is eight dirhems and four-fifths of thing. Subtract also the one dirhem which he has bequeathed; there remain seven dirhems and four-fifths of thing. Divide this between the two sons; there will be for each of them three dirhems and a half plus two-fifths of thing; and this is equal to one thing.* Reduce it by subtracting two-fifths of thing from thing. Then you have three-fifths of thing, equal to three dirhems and a half. Complete the thing by adding to it two-thirds of the same: add as much to the three dirhems and a half,

---

* $\frac{1}{2}[10+x]-1=2x$ $\therefore \frac{2}{5}[10+x]-\frac{1}{2}=x$

    $\therefore 3\frac{1}{2}=\frac{3}{5}x$ $\therefore x=\frac{35}{6}=5\frac{5}{6}$

The stranger receives $\frac{1}{5}[10+\frac{35}{6}]+1=4\frac{1}{6}$.

namely, two dirhems and one-third; the sum is five and five-sixths. This is the thing, or the amount which is taken from the debt.

If he leaves three sons, and bequeaths one-fifth of his property less one dirhem, leaving ten dirhems of capital and a demand of ten dirhems against one of the sons, the computation is this: You call the sum which is taken from the debt thing. Add this to the capital; it gives ten and thing. Subtract from this one-fifth of it for the legacy: it is two dirhems and one-fifth of thing. There remain eight dirhems and four-fifths of thing; add to this one dirhem, since he stated "less one dirhem." Thus you have nine dirhems and four-fifths of thing. Divide this between the three sons. There will be for each son three dirhems, and one-fifth and one-third and one-fifth of thing. This equals one thing.* Subtract one-fifth and one-third of one-

(67) fifth of thing from thing. There remain eleven-fifteenths of thing, equal to three dirhems. It is now required to complete the thing. For this purpose, add to it four-elevenths, and do the same with the three dirhems, by adding to them one dirhem and one-eleventh. Then you have four dirhems and one-eleventh, which are equal to thing. This is the sum which is taken out of the debt.

---

* $\frac{1}{5}[10+x]+1=3x$ ∴ $9=2\frac{1}{3}x$ ∴ $\frac{45}{11}$ or $4\frac{1}{11}=x$

The stranger receives $\frac{1}{5}[10+\frac{45}{11}]-1=1\frac{9}{11}$

On another Species of Legacy.

" A man dies, leaving his mother, his wife, and two brothers and two sisters by the same father and mother with himself; and he bequeaths to a stranger one-ninth of his capital."

Computation:* You constitute their shares by taking them out of forty-eight parts. You know that if you take one-ninth from any capital, eight-ninths of it will remain. Add now to the eight-ninths one-eighth of the same, and to the forty-eight also one-eighth of them, namely, six, in order to complete your capital. This gives fifty-four. The person to whom one-ninth is bequeathed receives six out of this, being one-ninth of the whole capital. The remaining forty-eight will be distributed among the heirs, proportionably to their legal shares.

If the instance be: "A woman dies, leaving her husband, a son, and three daughters, and bequeathing

---

* It appears in the sequel (p. 96) that a widow is entitled to $\frac{1}{8}$th, and a mother to $\frac{1}{6}$th of the residue; $\frac{1}{8}+\frac{1}{6}=\frac{14}{48}$, leaving $\frac{34}{48}$ of the residue to be distributed between two brothers and two sisters; that is, $\frac{17}{48}$ between a brother and a sister; but in what proportion these 17 parts are to be divided between the brother and sister does not appear in the course of this treatise.

Let the whole capital of the testator $= 1$
and let each 48th share of the residue $= x$

$$\frac{8}{9}=48x \quad \therefore \quad \frac{1}{9}=6x \quad \therefore \quad \frac{1}{54}=x$$

that is, each 48th part of the residue $= \frac{1}{54}$th of the whole capital.

to a stranger one-eighth and one-seventh of her capi-
(68) tal;" then you constitute the shares of the heirs, by
taking them out of twenty.* Take a capital, and subtract from it one-eighth and one-seventh of the same.
The remainder is, a capital less one-eighth and one-
seventh. Complete your capital by adding to that
which you have already, fifteen forty-one parts. Multiply the parts of the capital, which are twenty, by
forty-one; the product is eight hundred and twenty.
Add to it fifteen forty-one parts of the same, which are
three hundred: the sum is one thousand one hundred
and twenty parts. The person to whom one-eighth
and one-seventh were bequeathed, receives one-eighth
and one-seventh of this. One seventh of it is one hundred and sixty, and one-eighth one hundred and forty.
Subtracting this, there remain eight hundred and
twenty parts for the heirs, proportionably to their legal
shares.

---

* A husband is entitled to $\frac{1}{4}$th of the residue, and the
sons and daughters divide the remaining $\frac{3}{4}$ths of the residue
in such proportion, that a son receives twice as much as a
daughter. In the present instance, as there are three daughters
and one son, each daughter receives $\frac{1}{5}$ of $\frac{3}{4}, = \frac{3}{20}$, of the
residue, and the son, $\frac{6}{20}$. Since the stranger takes $\frac{1}{8} + \frac{1}{7} =
\frac{15}{56}$ of the capital, the residue $= \frac{41}{56}$ of the capital, and each
$\frac{1}{20}$th share of the residue $= \frac{1}{20} \times \frac{41}{56} = \frac{41}{1120}$ of the capital.
The stranger, therefore, receives $\frac{15}{56} = \frac{15 \times 20}{56 \times 20} = \frac{300}{1120}$ of the
capital.

On another Species of Legacies,* viz.

If nothing has been imposed on some of the heirs,† and something has been imposed on others; the legacy amounting to more than one-third. It must be known, that the law for such a case is, that if more than one-third of the legacy has been imposed on one of the heirs, this enters into his share; but that also those on whom nothing has been imposed must, nevertheless, contribute one-third.

Example: " A woman dies, leaving her husband, a son, and her mother. She bequeaths to a person two-fifths, and to another one-fourth of her capital. She imposes the two legacies together on her son, and on her mother one moiety (of the mother's share of the residue); on her husband she imposes nothing but one-third, (which he must contribute, according to the

---

* The problems in this chapter may be considered as belonging rather to Law than to Algebra, as they contain little more than enunciations of the law of inheritance in certain complicated cases.

† If some heirs are, by a testator, charged with payment of bequests, and other heirs are not charged with payment of any bequests whatever : if one bequest exceeds in amount $\frac{1}{3}$d of the testator's whole property ; and if one of his heirs is charged with payment of more than $\frac{1}{3}$d of such bequest ; then, whatever share of the residue such heir is entitled to receive, the like share must he pay of the bequest wherewith he is charged, and those heirs whom the testator has not charged with any payment, must each contribute towards paying the bequests a third part of their several shares of the residue.

law)."* Computation: You constitute the shares of the
(69) heritage, by taking them out of twelve parts: the son
receives seven of them, the husband three, and the
mother two parts. You know that the husband must
give up one-third of his share; accordingly he retains
twice as much as that which is detracted from his share
for the legacy. As he has three parts in hand, one of
these falls to the legacy, and the remaining two parts
he retains for himself. The two legacies together are
imposed upon the son. It is therefore necessary to
subtract from his share two-fifths and one-fourth of the
same. He thus retains seven twentieths of his entire
original share, dividing the whole of it into twenty
equal parts. The mother retains as much as she contributes to the legacy; this is one (twelfth part), the
entire amount of what she had received being two parts.

---

* If the bequests stated in the present example were charged
on the heirs collectively, the husband would be entitled to $\frac{1}{4}$,
the mother to $\frac{1}{6}$ of the residue : $\frac{1}{4}+\frac{1}{6}=\frac{5}{12}$; the remainder $\frac{7}{12}$
would be the son's share of the residue; but since the
bequests, $\frac{2}{5}+\frac{1}{4}=\frac{13}{20}$ of the capital, are charged upon the son
and mother, the law throws a portion of the charge on the
husband.

The Husband contributes $\frac{1}{4} \times \frac{1}{3} = 20 \times \frac{1}{240}$, and retains $\frac{1}{4} \times \frac{2}{3} = 40 \times \frac{1}{240}$

The Mother . . . . $\frac{1}{6} \times \frac{1}{2} = 20 \times \frac{1}{240}$, ... $\frac{1}{6} \times \frac{1}{2} = 20 \times \frac{1}{240}$

The Son . . . . $\frac{7}{12} \times \frac{13}{20} = 91 \times \frac{1}{240}$, ... $\frac{7}{12} \times \frac{7}{20} = 49 \times \frac{1}{240}$

$$\text{Total contributed} = \frac{131}{240} \qquad \text{Total retained} = \frac{109}{240}$$
$$\frac{2}{5}+\frac{1}{4}=\frac{8}{20}+\frac{5}{20}=\frac{13}{20}$$

The Legatee, to whom the $\frac{2}{5}$ are bequeathed, receives $\frac{8}{13} \times \frac{131}{240} = \frac{8 \times 131}{3120}$

The Legatee, to whom $\frac{1}{4}$ is bequeathed, receives $\frac{5}{13} \times \frac{131}{240} = \frac{5 \times 131}{3120}$

Take now a sum, one-fourth of which may be divided into thirds, or of one-sixth of which the moiety may be taken; this being again divisible by twenty. Such a capital is two hundred and forty. The mother receives one-sixth of this, namely, forty; twenty from this fall to the legacy, and she retains twenty for herself. The husband receives one-fourth, namely, sixty; from which twenty belong to the legacy, so that he retains forty. The remaining hundred and forty belong to the son; the legacy from this is two-fifths and one-fourth, or ninety-one; so that there remain forty-nine. The entire sum for the legacies is, therefore, one hundred and thirty-one, which must be divided among the two legatees. The one to whom two-fifths were bequeathed, receives eight-thirteenths of this; the one to whom one-fourth was devised, receives five-thirteenths. If you wish distinctly to express the shares of the two legatees, you need only to multiply (70) the parts of the heritage by thirteen, and to take them out of a capital of three thousand one hundred and twenty.

But if she had imposed on her son (payment of) the two-fifths to the person to whom the two-fifths were bequeathed, and of nothing to the other legatee; and upon her mother (payment of) the one-fourth to the person to whom one-fourth was granted, and of nothing to the other legatee; and upon her husband nothing besides the one-third (which he must according to law contribute) to both; then you know that this one-third comes to the advantage of the heirs collectively; and the legatee of the two-fifths receives eight-thirteenths, and the legatee of the one-fourth receives five-thirteenths from it. Constitute the shares as I have shown above, by taking twelve parts; the husband receives one-fourth of them, the mother one-sixth, and the son that which remains.* Computation: You know that at all events the husband must give up one-third of his share, which consists of three parts. The mother must likewise give up one-third, of which each legatee partakes according to the proportion of his legacy. Besides, she must pay to the legatee to whom one-fourth is bequeathed, and whose legacy has been imposed on her, as much as the difference between the one-fourth and his

---

* $\frac{2}{5} + \frac{1}{4} = \frac{8+5}{20} = \frac{13}{20}$

The Husband, who would be entitled to $\frac{1}{4}$ of the residue, is not charged by the Testator with any bequest.

The Mother who would be entitled to $\frac{1}{6}$ of the residue, is charged with the payment of $\frac{1}{4}$ to the Legatee A.

The Son, who would be entitled to $\frac{7}{12}$ of the residue, is charged with payment of $\frac{2}{5}$ to the Legatee B.

The Husband contributes $\}$ $\frac{1}{4} \times \frac{1}{3} = 780 \times \frac{1}{9360}$; retains $\frac{1}{4} \times \frac{2}{3} = \frac{1560}{9360}$

The Mother .... $\frac{1}{6}\left[\frac{1}{4} + \frac{8}{13} \times \frac{1}{3}\right] = 710 \times \frac{1}{9360}$; retains ...... $\frac{850}{9360}$

The Son ..... $\frac{7}{12}\left[\frac{2}{5} + \frac{5}{13} \times \frac{1}{3}\right] = 2884 \times \frac{1}{9360}$; retains ...... $\frac{2576}{9360}$

Total contributed $= \frac{4374}{9360}$; Total retained $= \frac{4986}{9360}$

The Legatee A, to whom $\frac{1}{4}$ is bequeathed, receives $\}$ $\frac{5}{13} \times \frac{4374}{9360} = \frac{5 \times 4374}{964080}$

The Legatee B, to whom $\frac{2}{5}$ are bequeathed, receives $\}$ $\frac{8}{13} \times \frac{4374}{9360} = \frac{8 \times 4374}{964080}$

portion of the one-third, namely, nineteen one hundred and fifty-sixths of her entire share, considering her share as consisting of one hundred and fifty-six parts. His portion of the one-third of her share is twenty parts. But what she gives him is one-fourth of her entire share, namely, thirty-nine parts. One third of her share is taken for both legacies, and besides nineteen parts which she must pay to him alone. The son gives to the legatee to whom two-fifths are bequeathed as much as the difference between two-fifths of his (the son's) share (71) and the legatee's portion of the one-third, namely, thirty-eight one hundred and ninety-fifths of his (the son's) entire share, besides the one-third of it which is taken off from both legacies. The portion which he (the legatee) receives from this one-third, is eight-thirteenths of it, namely, forty (one hundred and ninety-fifths); and what the son contributes of the two-fifths from his share is thirty-eight. These together make seventy-eight. Consequently, sixty-five will be taken from the son, as being one-third of his share, for both legacies, and besides this he gives thirty-eight to the one of them in particular. If you wish to express the parts of the heritage distinctly, you may do so with nine hundred and sixty-four thousand and eighty.

On another Species of Legacies.

" A man dies, leaving four sons and his wife; and bequeathing to a person as much as the share of one

of the sons less the amount of the share of the widow."
Divide the heritage into thirty-two parts. The widow
receives one-eighth,* namely, four; and each son seven.
Consequently the legatee must receive three-sevenths of
the share of a son. Add, therefore, to the heritage
three-sevenths of the share of a son, that is to say,
three parts, which is the amount of the legacy. This
gives thirty-five, from which the legatee receives three;
and the remaining thirty-two are distributed among
the heirs proportionably to their legal shares.

If he leaves two sons and a daughter,† and bequeaths
to some one as much as would be the share of a third
son, if he had one; then you must consider, what
(72) would be the share of each son, in case he had three.
Assume this to be seven, and for the entire heritage

---

* A widow is entitled to $\frac{1}{8}$th of the residue; therefore
$\frac{7}{8}$ths of the residue are to be distributed among the sons of
the testator. Let x be the stranger's legacy. The widow's
share $= \frac{1-x}{8}$; each son's share $= \frac{1}{4} \times \frac{7}{8} [1-x]$; and a son's
share, minus the widow's share $= [\frac{7}{4} - 1] \frac{1-x}{8} = \frac{3}{4} \cdot \frac{1-x}{8}$
$\therefore x = \frac{3}{4} \cdot \frac{1-x}{8}$ $\therefore x = \frac{3}{35}$; $1-x = \frac{32}{35}$ A son's share $= \frac{7}{35}$;
the widow's share $= \frac{4}{35}$.

† A son is entitled to receive twice as much as a daughter.
Were there three sons and one daughter, each son would
receive $\frac{2}{7}$ths of the residue. Let x be the stranger's legacy.
$\therefore \frac{2}{7} [1-x] = x$ $\therefore x = \frac{2}{9}$, and $1-x = \frac{7}{9}$
Each Son's share.... $= \frac{2}{9} [1-x] = \frac{2}{9} \times \frac{7}{9} = \frac{14}{45}$
The Daughter's share $= \frac{1}{9} [1-x]$ ...... $= \frac{7}{45}$
The Stranger's legacy $= \frac{2}{9}$ .... $= \frac{10}{45}$

take a number, one-fifth of which may be divided into sevenths, and one-seventh of which may be divided into fifths. Such a number is thirty-five. Add to it two-sevenths of the same, namely, ten. This gives forty-five. Herefrom the legatee receives ten, each son fourteen, and the daughter seven.

If he leaves a mother, three sons, and a daughter, and bequeaths to some one as much as the share of one of his sons less the amount of the share of a second daughter, in case he had one; then you distribute the heritage into such a number of parts as may be divided among the actual heirs, and also among the same, if a second daughter were added to them.* Such a number is three hundred and thirty-six. The share of the second daughter, if there were one, would be thirty-five, and that of a son eighty: their difference is forty-five, and this is the legacy. Add to it three hundred and thirty-six, the sum is three hundred and eighty-one, which is the number of parts of the entire heritage.

---

* Let x be the stranger's legacy; $1-x$ is the residue.

A widow's share of the residue is $\frac{1}{6}$th: there remains $\frac{5}{6}[1-x]$, to be distributed among the children.

Since there are 3 sons, and 1 daughter, a son's share is .... $\Big\} \frac{2}{7} \times \frac{5}{6}[1-x]$

Were there 3 sons and 2 daughters, a daughter's share would be ...... $\Big\} \frac{1}{8} \times \frac{5}{6}[1-x]$

The difference $= \frac{9}{56} \times \frac{5}{6}[1-x]$

$\therefore x = \frac{45}{336}[1-x] \qquad \therefore x = \frac{45}{381}$

$1-x = \frac{336}{381}$; the widow's share $= \frac{56}{381}$

the daughter's share $= \frac{40}{381}$

If he leaves three sons, and bequeaths to some one as much as the share of one of his sons, less the share of a daughter, supposing he had one, plus one-third of the remainder of the one-third; the computation will be this :* distribute the heritage into such a number of parts as may be divided among the actual heirs, and also among them if a daughter were added to them. Such a number is twenty-one. Were a daughter among the heirs, her share would be three, and that of a son seven. The testator has therefore bequeathed to the (73) legatee four-sevenths of the share of a son, and one-third of what remains from one-third. Take therefore one-third, and remove from it four-sevenths of the share of a son. There remains one-third of the capital less four-sevenths of the share of a son. Subtract now one-third of what remains of the one-third, that is to say, one-ninth of the capital less one-seventh and one-third of the seventh of the share of a son; the remainder

---

* Since there are 3 sons, each son's share of the residue $=\frac{1}{3}$. Were there 3 sons and a daughter, the daughter's share would be $\frac{1}{7}$.

$$\tfrac{1}{3}-\tfrac{1}{7}=\tfrac{4}{7}$$

Let x be the stranger's legacy, and v a son's share

Then $1-x=3v$

but $x=\tfrac{4}{7}v+\tfrac{1}{3}\left[\tfrac{1}{3}-\tfrac{4}{7}v\right]$

and $1-x=\tfrac{2}{3}+\tfrac{1}{3}-\tfrac{4}{7}v-\tfrac{1}{3}\left[\tfrac{1}{3}-\tfrac{4}{7}v\right]=3v$

$\therefore \tfrac{2}{3}+\tfrac{2}{3}\left[\tfrac{1}{3}-\tfrac{4}{7}v\right]=3v$

$\therefore \tfrac{2}{3}+\tfrac{2}{9}=3\tfrac{8}{21}\times v$, or $\tfrac{8}{9}=\tfrac{71}{21}v$

$\therefore \tfrac{8}{3}=\tfrac{71}{7}v$ $\therefore v=\tfrac{56}{213}=$ a son's share

$x=\tfrac{45}{213}=$ the stranger's legacy.

is two-ninths of the capital less two-sevenths and two-thirds of a seventh of the share of a son. Add this to the two-thirds of the capital; the sum is eight-ninths of the capital less two-sevenths and two thirds of a seventh of the share of a son, or eight twenty-one parts of that share, and this is equal to three shares. Reduce this, you have then eight-ninths of the capital, equal to three shares and eight twenty-one parts of a share. Complete the capital by adding to eight-ninths as much as one-eighth of the same, and add in the same proportion to the shares. Then you find the capital equal to three shares and forty-five fifty-sixth parts of a share. Calculating now each share equal to fifty-six, the whole capital is two hundred and thirteen, the first legacy thirty-two, the second thirteen, and of the remaining one hundred and sixty-eight each son takes fifty-six.

———

On another Species of Legacies.

" A woman dies, leaving her daughter, her mother, and her husband, and bequeaths to some one as much as the share of her mother, and to another as much as one-ninth of her entire capital."* Computation: You begin by dividing the heritage into thirteen parts, two

———

* In the former examples (p. 90) when a husband and a mother were among the heirs, a husband was found to be entitled to $\frac{1}{4}=\frac{3}{12}$, and a mother to $\frac{1}{6}=\frac{2}{12}$ of the residue. Here a husband is stated to be entitled to $\frac{3}{13}$, and a mother to $\frac{2}{13}$ of the residue.

of which the mother receives. Now you perceive that the
(74) legacies amount to two parts plus one-ninth of the entire capital. Subtracting this, there remains eight-ninths
of the capital less two parts, for distribution among
the heirs. Complete the capital, by making the eight-
ninths less two parts to be thirteen parts, and adding
two parts to it, so that you have fifteen parts, equal
to eight-ninths of capital; then add to this one-
eighth of the same, and to the fifteen parts add likewise one-eighth of the same, namely, one part and
seven-eighths; then you have sixteen parts and seven-
eighths. The person to whom one-ninth is bequeathed,
receives one-ninth of this, namely, one part and seven-
eighths; the other, to whom as much as the share of
the mother is bequeathed, receives two parts. The
remaining thirteen parts are divided among the heirs,
according to their legal shares. You best determine
the respective shares by dividing the whole heritage
into one hundred and thirty-five parts.

If she has bequeathed as much as the share of the
husband and one-eighth and one-tenth of the capital,*

Let $\frac{1}{13}$ of the residue $= v$

$1 - \frac{1}{9} - 2v = 13v$ $\therefore \frac{8}{9} = 15v$

$\therefore v = \frac{8}{135}$ of the capital

A mother's share $= \frac{16}{135}$

* $\frac{1}{8} + \frac{1}{10} = \frac{9}{40}$

A husband's share of the residue is $\frac{3}{13}$

$\therefore 1 - \frac{9}{40} - 3v = 13v$ $\therefore \frac{31}{40} = 16v$

$\therefore v = \frac{31}{640}$; a husband's share $= \frac{93}{640}$

The stranger's legacy $= \frac{237}{640}$

then you begin by dividing the heritage into thirteen parts. Add to this as much as the share of the husband, namely, three; thus you have sixteen. This is what remains of the capital after the deduction of one-eighth and one-tenth, that is to say, of nine-fortieths. The remainder of the capital, after the deduction of one-eighth and one-tenth, is thirty-one fortieths of the same, which must be equal to sixteen parts. Complete your capital by adding to it nine thirty-one parts of the same, and multiply sixteen by thirty-one, which gives four hundred and ninety-six; add to this nine thirty-one parts of the same, which is one hundred and forty- (75) four. The sum is six hundred and forty. Subtract one-eighth and one-tenth from it, which is one hundred and forty-four, and as much as the share of the husband, which is ninety-three. There remains four hundred and three, of which the husband receives ninety-three, the mother sixty-two, and every daughter one hundred and twenty-four.

If the heirs are the same,* but that she bequeaths to a person as much as the share of the husband, less one-ninth and one-tenth of what remains of the capital,

---

$$* \quad \tfrac{1}{9} + \tfrac{1}{10} = \tfrac{19}{90}$$

$$1 - 3v + \tfrac{19}{90}\left[1 - 3v\right] = 13v$$

$$\therefore \tfrac{109}{90}\left[1 - 3\right] = 13v$$

$$\therefore \tfrac{109}{90} = \left[13 + \tfrac{109}{30}\right]v$$

$$\therefore v = \tfrac{109}{1497}$$

The husband's share $= \tfrac{327}{1497}$

The stranger's legacy $= \tfrac{80}{1497}$

after the subtraction of that share, the computation is this: Divide the heritage into thirteen parts. The legacy from the whole capital is three parts, after the subtraction of which there remains the capital less three parts. Now, one-ninth and one-tenth of the remaining capital must be added, namely, one-ninth and one-tenth of the whole capital less one-ninth and one-tenth of three parts, or less nineteen-thirtieths of a part; this yields the capital and one-ninth and one-tenth less three parts and nineteen-thirtieths of a part, equal to thirteen parts. Reduce this, by removing the three parts and nineteen-thirtieths from your capital, and adding them to the thirteen parts. Then you have the capital and one-ninth and one-tenth of the same, equal to sixteen parts and nineteen-thirtieths of a part. Reduce this to one capital, by subtracting from it nineteen one-hundred-and-ninths. There remains a (76) capital, equal to thirteen parts and eighty one-hundred-and-ninths. Divide each part into one hundred and nine parts, by multiplying thirteen by one hundred and nine, and add eighty to it. This gives one thousand four hundred and ninety-seven parts. The share of the husband from it is three hundred and twenty-seven parts.

If some one leaves two sisters and a wife,* and bequeaths to another person as much as the share of a

---

* When the heirs are a wife, and 2 sisters, they each inherit $\frac{1}{3}$ of the residue.

Let

sister less one-eighth of what remains of the capital after the deduction of the legacy, the computation is this : You consider the heritage as consisting of twelve parts. Each sister receives one-third of what remains of the capital after the subtraction of the legacy; that is, of the capital less the legacy. You perceive that one-eighth of the remainder plus the legacy equals the share of a sister; and also, one-eighth of the remainder is as much as one-eighth of the whole capital less one-eighth of the legacy; and again, one-eighth of the capital less one-eighth of the legacy added to the legacy equals the share of a sister, namely, one-eighth of the capital and seven-eighths of the legacy. The whole capital is therefore equal to three-eighths of the capital plus three and five-eighth times the legacy. Subtract now from the capital three-eighths of the same. There remain five-eighths of the capital, equal to three and five-eighth times the legacy; and the entire capital is equal to five and four-fifth times the legacy. Consequently, if you assume the capital to be twenty-nine, the legacy is five, and each sister's share eight.

Let x be the stranger's legacy.

$\frac{1}{3}[1-x] =$ a sister's share

$\frac{1}{3}[1-x] - \frac{1}{8}[1-x] = x$

$\therefore \frac{5}{24}[1-x] = x \quad \therefore \frac{5}{24} = \frac{29}{24}x$

$\therefore x = \frac{5}{29} \quad \therefore 1-x = \frac{24}{29}$

and a sister's share $= \frac{8}{29}$

On another Species of Legacies.

" A man dies, and leaves four sons, and bequeaths to some person as much as the share of one of his sons; and to another, one-fourth of what remains after the deduction of the above share from one-third." You perceive that this legacy belongs to the class of those (77) which are taken from one-third of the capital.* Computation : Take one-third of the capital, and subtract from it the share of a son. The remainder is one-third of the capital less the share. Then subtract from it one-fourth of what remains of the one-third, namely, one-fourth of one-third less one-fourth of the share. The remainder is one-fourth of the capital less three-fourths of the share. Add hereto two-thirds of the capital : then you have eleven-twelfths of the capital less three-fourths of a share, equal to four shares. Reduce this by removing the three-fourths of the share from the capital, and adding them to the four shares. Then you have eleven-twelfths of the capital, equal to four shares and three-fourths. Complete your capital, by adding to the four shares and three-fourths one-fourth of the same. Then you have five shares and two-elevenths,

---

* Let the first bequest $=v$; and the second $=y$

Then $1-v-y=4v$

i.e. $\frac{2}{3}+\frac{1}{3}-v-\frac{1}{4}\left[\frac{1}{3}-v\right]=4v$

$\therefore \frac{2}{3}+\frac{3}{4}\left[\frac{1}{3}-v\right]=4v$

$\therefore \frac{2}{3}+\frac{3}{12}=\left[4+\frac{3}{4}\right]v \quad \therefore \frac{11}{12}=\frac{19}{4}v$

$\therefore v=\frac{11}{37}$; the 2d bequest $=\frac{2}{37}$

equal to the capital. Suppose, now, every share to be eleven; then the whole square will be fifty-seven; one-third of this is nineteen; from this one share, namely, eleven, must be subtracted; there remain eight. The legatee, to whom one-fourth of this remainder was bequeathed, receives two. The remaining six are returned to the other two-thirds, which are thirty-eight. Their sum is forty-four, which is to be divided amongst the four sons; so that each son receives eleven.

If he leaves four sons, and bequeaths to a person as much as the share of a son, less one-fifth of what remains from one-third after the deduction of that share, then this is likewise a legacy, which is taken from one-third.* Take one-third, and subtract from it one share; there remains one-third less the share. Then return to it that which was excepted, namely, one-fifth of the one-third less one-fifth of the share. This gives one-third and one-fifth of one-third (or two-fifths) (78) less one share and one-fifth of a share. Add this to two-thirds of the capital. The sum is, the capital and one-third of one-fifth of the capital less one share and one-fifth of a share, equal to four shares. Reduce this by removing one share and one-fifth from the capital,

---

* $1 - v + \frac{1}{5}\left[\frac{1}{3} - v\right] - 4v$

or $\frac{2}{3} \times \frac{1}{3} - v + \frac{1}{5}\left[\frac{1}{3} - v\right] = 4v$

or $\frac{2}{3} + \frac{6}{5}\left[\frac{1}{3} - v\right] = 4v$

$\therefore \frac{2}{3} + \frac{2}{5} = \left[4 + \frac{6}{5}\right]v \quad \therefore \frac{16}{15} = \frac{26}{5}v$

$\therefore v = \frac{8}{39}$, and the stranger's legacy $= \frac{7}{39}$

and add to it the four shares. Then you have the capital and one-third of one-fifth of the capital, which are equal to five shares and one-fifth. Reduce this to one capital, by subtracting from what you have the moiety of one-eighth of it, that is to say, one-sixteenth. Then you find the capital equal to four shares and seven-eighths of a share. Assume now thirty-nine as capital; one-third of it will be thirteen, and one share eight; what remains of one-third, after the deduction of that share, is five, and one-fifth of this is one. Subtract now the one, which was excepted from the legacy; the remaining legacy then is seven; subtracting this from the one-third of the capital, there remain six. Add this to the two-thirds of the capital, namely, to the twenty-six parts, the sum is thirty-two; which, when distributed among the four sons, yields eight for each of them.

If he leaves three sons and a daughter,* and bequeaths to some person as much as the share of a

---

* Since there are three sons and one daughter, the daughter receives $\frac{1}{7}$, and each son $\frac{2}{7}$ths of the residue.

If the 1st legacy $= v$, the 2d $= y$, and therefore a daughter's share $= v$,

$$1 - v - y = 7v; \quad \tfrac{1}{5} + \tfrac{1}{6} = \tfrac{11}{30}$$

$$\therefore \tfrac{5}{7} + \tfrac{2}{7} - v - \tfrac{11}{30} \left[\tfrac{2}{7} - v \right] = 7v$$

$$\text{i.e. } \tfrac{5}{7} + \tfrac{19}{30} \left[\tfrac{2}{7} - v \right] = 7v$$

$$\therefore \tfrac{5}{7} + \tfrac{19}{15 \times 7} = \left[7 + \tfrac{19}{30} \right] v$$

$$\therefore \tfrac{94}{7} = \tfrac{229}{9} v \quad \therefore = \tfrac{188}{1603}$$

The 2d legacy $= \therefore y = \tfrac{20}{1603}$

daughter, and to another one-fifth and one-sixth of what remains of two-sevenths of the capital after the deduction of the first legacy; then this legacy is to be taken out of two-sevenths of the capital. Subtract from two-sevenths the share of the daughter: there remain two-sevenths of the capital less that share. Deduct from this the second legacy, which comprises (79) one-fifth and one-sixth of this remainder: there remain one-seventh and four-fifteenths of one-seventh of the capital less nineteen-thirtieths of the share. Add to this the other five-sevenths of the capital: then you have six-sevenths and four-fifteenths of one-seventh of the capital less nineteen thirtieths of the share, equal to seven shares. Reduce this, by removing the nineteen thirtieths, and adding them to the seven shares: then you have six-sevenths and four-fifteenths of one-seventh of capital, equal to seven shares and nineteen-thirtieths. Complete your capital by adding to every thing that you have eleven ninety-fourths of the same; thus the capital will be equal to eight shares and ninety-nine one hundred and eighty-eighths. Assume now the capital to be one thousand six hundred and three; then the share of the daughter is one hundred and eighty-eight. Take two-sevenths of the capital; that is, four hundred and fifty-eight. Subtract from this the share, which is one hundred and eighty-eight; there remain two hundred and seventy. Remove one-fifth and one-sixth of this, namely, ninety-nine; the remainder is one hundred and seventy-one. Add thereto five-

sevenths of the capital, which is one thousand one
hundred and forty-five. The sum is one thousand three
(80) hundred and sixteen parts. This may be divided into
seven shares, each of one hundred and eighty-eight
parts ; then this is the share of the daughter, whilst
every son receives twice as much.

If the heirs are the same, and he bequeaths to some
person as much as the share of the daughter, and to
another person one-fourth and one-fifth out of what
remains from two-fifths of his capital after the deduction of the share; this is the computation :* You must
observe that the legacy is determined by the two-fifths.
Take two-fifths of the capital and subtract the shares :
the remainder is, two-fifths of the capital less the share.
Subtract from this remainder one-fourth and one-fifth
of the same, namely, nine-twentieths of two-fifths, less
as much of the share. The remainder is one-fifth
and one-tenth of one-fifth of the capital less eleven-
twentieths of the share. Add thereto three-fifths of the

$$* \ \tfrac{1}{4} + \tfrac{1}{5} = \tfrac{9}{20}$$

Let the 1st legacy $= v =$ a daughter's share

Let the 2d legacy $= y$

$$1 - v - y = 7v$$

$$\therefore \tfrac{2}{5} + \tfrac{2}{5} - v - \tfrac{9}{20} \left[\tfrac{2}{5} - v \right] = 7v$$

$$\therefore \tfrac{3}{5} + \tfrac{11}{20} \left[\tfrac{2}{5} - v \right] = 7v$$

$$\therefore \tfrac{3}{5} + \tfrac{11}{10 \times 5} = \left[7 + \tfrac{11}{20} \right] v$$

$$\therefore \tfrac{41}{5} = \tfrac{151}{2} v \quad \therefore v = \tfrac{82}{755}$$

and the 2d legacy, $y, = \tfrac{99}{155}$

capital: the sum is four-fifths and one-tenth of one-fifth of the capital, less eleven-twentieths of the share, equal to seven shares. Reduce this by removing the eleven-twentieths of a share, and adding them to the seven shares. Then you have the same four-fifths and one-tenth of one-fifth of capital, equal to seven shares and eleven-twentieths. Complete the capital by adding to any thing that you have nine forty-one parts. Then you have capital equal to nine shares and seventeen eighty-seconds. Now assume each portion to consist of eighty-two parts; then you have seven hundred and fifty-five parts. Two-fifths of these are three hundred (81) and two. Subtract from this the share of the daughter, which is eighty-two; there remain two hundred and twenty. Subtract from this one-fourth and one-fifth, namely, ninety-nine parts. There remain one hundred and twenty-one. Add to this three-fifths of the capital, namely, four hundred and fifty-three. Then you have five hundred and seventy-four, to be divided into seven shares, each of eighty-two parts. This is the share of the daughter; each son receives twice as much.

If the heirs are the same, and he bequeaths to a person as much as the share of a son, less one-fourth and one-fifth of what remains of two-fifths (of the capital) after the deduction of the share; then you see that this legacy is likewise determined by two-fifths. Subtract two shares (of a daughter) from them, since every son receives two (such) shares; there remain two-fifths of the capital less two (such) shares. Add thereto what was excepted from the legacy, namely, one-fourth and one-fifth of the two-fifths less nine-tenths of (a daughter's) share.* Then you have two-fifths and nine-tenths of one-fifth of the capital less two (daughter's) shares and nine-tenths. Add to this three-fifths of the capital. Then you have one capital and nine-tenths of one-fifth of the capital less two (daughter's) shares and nine-tenths, equal to seven (such) shares. Reduce this by removing the two shares and nine-tenths and adding them to the seven shares. Then you have one capital and nine-tenths of one-fifth of the capital, equal to nine shares of a daughter and nine-tenths. (82) Reduce this to one entire capital, by deducting nine fifty-ninths from what you have. There remains the capital equal to eight such shares and twenty-three fifty-ninths. Assume now each share (of a daughter) to contain fifty-nine parts. Then the whole heritage comprizes four hundred and ninety-five parts. Two-fifths of this are one hundred and ninety-eight

---

* $v = \frac{1}{7}$ of the residue = a daughter's share.

$$2v = \text{a son's share}$$

$$1 - 2v + \frac{9}{20}\left[\frac{2}{5} - 2v\right] = 7v$$

i.e. $\frac{3}{5} + \frac{2}{5} - 2v + \frac{9}{20}\left[\frac{2}{5} - 2v\right] = 7v$

$\therefore \frac{3}{5} + \frac{29}{20}\left[\frac{2}{5} - 2v\right] = 7v$

$\therefore \frac{3}{5} + \frac{29}{10 \times} = \left[7 + \frac{29}{10}\right]v$ $\quad \therefore \frac{59}{5} = 99v$

$\therefore v = \frac{59}{495}$; a son's share $= \frac{118}{495}$

and the legacy to the stranger $= \frac{80}{495}$

parts. Subtract therefrom the two shares (of a daughter) or one hundred and eighteen parts; there remain eighty parts. Subtract now that which was excepted, namely, one-fourth and one fifth of these eighty, or thirty-six parts; there remain for the legatee eighty-two parts. Deduct this from the parts in the total number of parts in the heritage, namely, four hundred and ninety-five. There remain four hundred and thirteen parts to be distributed into seven shares; the daughter receiving (one share or) fifty-nine (parts), and each son twice as much.

If he leaves two sons and two daughters, and bequeaths to some person as much as the share* of a

---

* Since there are two sons and two daughters, each son receives $\frac{1}{3}$, and each daughter $\frac{1}{6}$ of the residue. Let $v = $ a daughter's share.

$$\text{Let the 1st legacy} = x = v - \frac{1}{5}\left[\frac{1}{3} - v\right]$$
$$\cdots\cdots\ \text{2d}\ \cdots\cdots = y = v - \frac{1}{3}\left[\frac{1}{3} - x - v\right]$$
$$\text{and 3d}\ \cdots\cdots = \frac{1}{12}$$
$$1 - \frac{1}{12} - x - y = 6v$$
$$\text{i.e.}\ \frac{2}{3} - \frac{1}{12} + \frac{1}{3} - x - v + \frac{1}{3}\left[\frac{1}{3} - x - v\right] = 6v$$
$$\text{or}\ \frac{2}{3} - \frac{1}{12} + \frac{4}{3}\left[\frac{1}{3} - x - v\right] = 6v$$
$$\text{i.e.}\ \frac{7}{12} + \frac{4}{3}\left[\frac{1}{3} - v + \frac{1}{5}\left[\frac{1}{3} - v\right] - v\right] = 6v$$
$$\text{or}\ \frac{7}{12} + \frac{4}{3}\left[\frac{6}{5}\left[\frac{1}{3} - v\right] - v\right] 6v$$
$$\text{or}\ \frac{7}{12} + \frac{8}{15} = \left[6 + \frac{4 \times 11}{3 \times 5}\right]v = \frac{134}{15}v$$
$$\text{or}\ \frac{7}{4} + \frac{8}{5} = \frac{134}{5}a\ \therefore v = \frac{67}{536} = \frac{1}{8}$$
$$\text{The 1st Legacy} = x = \frac{1}{12}$$
$$\text{The 2d}\ \cdots\cdots = y = \frac{1}{12}$$
$$\text{A son's share} = \frac{1}{4}$$

daughter less one-fifth of what remains from one-third after the deduction of that share; and to another person as much as the share of the other daughter less one-third of what remains from one-third after the deduction of all this; and to another person half one-sixth of his entire capital; then you observe that all these legacies are determined by the one-third. Take one-third of the capital, and subtract from it the share of a daughter; there remains one-third of the capital less one share. Add to this that which was excepted, namely, one-fifth of the one-third less one-fifth of the share: this gives one-third and one-fifth of one-third of

(83) the capital less one and one-fifth portion. Subtract herefrom the portion of the second daughter; there remain one-third and one-fifth of one-third of the capital less two portions and one-fifth. Add to this that which was excepted; then you have one-third and three-fifths of one-third, less two portions and fourteen-fifteenths of a portion. Subtract herefrom half one-sixth of the entire capital: there remain twenty-seven sixtieths of the capital less the two shares and fourteen-fifteenths, which are to be subtracted. Add thereto two-thirds of the capital, and reduce it, by removing the shares which are to be subtracted, and adding them to the other shares. You have then one and seven-sixtieths of capital, equal to eight shares and fourteen-fifteenths. Reduce this to one capital by subtracting from every thing that you have seven-sixtieths. Then let a share be two hundred

and one;* the whole capital will be one thousand six hundred and eight.

If the heirs are the same, and he bequeaths to a person as much as the share of a daughter, and one-fifth of what remains from one-third after the deduction of that share; and to another as much as the share of the second daughter and one-third of what remains from one-fourth after the deduction of that share: then, in the computation,† you must consider that the two legacies are determined by one-fourth and one-third. Take one-third of the capital, and subtract from it one share; there remains one-third of the capital less one share. Then subtract one-fifth of the remainder, namely, one-fifth of one-third of the capital, less one-fifth of the share; there remain four-fifths of one-third, less four-fifths of the share. Then take also one-fourth of the capital, and subtract from it one (84) share; there remains one-fourth of the capital, less one share. Subtract one-third of this remainder: there

---

* $\frac{201}{1608} = \frac{1}{8} = \frac{3}{24} = v$; and $\frac{1}{12} = \frac{2}{24} = y$

The common denominator 1608 is unnecessarily great.

† Let x be the 1st legacy; y the 2d; v a daughter's share.

$$1 - x - y = 6v$$
$$x = v + \frac{1}{3}\left[\frac{1}{3} - v\right]$$
$$y = v + \frac{1}{3}\left[\frac{1}{4} - v\right]$$

Then $1 - \frac{1}{3} - \frac{1}{4} + \frac{1}{3} - v - \frac{1}{5}\left[\frac{1}{3} - v\right] + \frac{1}{4} - v - \frac{1}{3}\left[\frac{1}{4} - v\right] = 6v$

or $\frac{5}{12} + \frac{4}{5}\left[\frac{1}{3} - v\right] + \frac{2}{3}\left[\frac{1}{4} - v\right] = 6v$

∴ $\frac{5}{12} + \frac{4}{15} + \frac{2}{12} = \left[6 + \frac{4}{5} + \frac{2}{3}\right]v$

∴ $\frac{51}{60} = \frac{112}{15}v$ ∴ $\frac{51}{448} = \frac{153}{1344}$

$$x = \frac{212}{1344}; \quad y = \frac{214}{1344}$$

Q

remain two-thirds of one-fourth of the capital, less two-thirds of one share. Add this to the remainder from the one-third of the capital; the sum will be twenty-six sixtieths of the capital, less one share and twenty-eight sixtieths. Add thereto as much as remains of the capital after the deduction of one-third and one-fourth from it; that is to say, one-fourth and one-sixth; the sum is seventeen-twentieths of the capital, equal to seven shares and seven-fifteenths. Complete the capital, by adding to the portions which you have three-seventeenths of the same. Then you have one capital, equal to eight shares and one-hundred-and-twenty hundred-and-fifty-thirds. Assume now one share to consist of one-hundred-and-fifty-three parts, then the capital consists of one thousand three hundred and forty-four. The legacy determined by one-third, after the deduction of one share, is fifty-nine; and the legacy determined by one-fourth, after the deduction of the share, is sixty-one.

If he leaves six sons, and bequeaths to a person as much as the share of a son and one-fifth of what remains of one-fourth; and to another person as much as the share of another son less one-fourth of what remains of one-third, after the deduction of the two first legacies and the second share; the computation is this :*
You subtract one share from one-fourth of the capital;

---

* Let x be the legacy to the 1st stranger
   and y .... 2d ......; $v =$ a son's share

there remains one-fourth less the share. Remove then (85) one-fifth of what remains of the one-fourth, namely, half one-tenth of the capital less one-fifth of the share. Then return to the one-third, and deduct from it half one-tenth of the capital, and four-fifths of a share, and one other share besides. The remainder then is one-third, less half one-tenth of the capital, and less one share and four-fifths. Add hereto one-fourth of the remainder, which was excepted, and assume the one-third to be eighty; subtracting from it half one-tenth of the capital, there remain of it sixty-eight less one share and four-fifths. Add to this one-fourth of it, namely, seventeen parts, less one-fourth of the shares to be subtracted from the parts. Then you have eighty-five parts less two shares and one-fourth. Add this to the other two-thirds of the capital, namely, one hundred and sixty parts. Then you have one and one-eighth of one-sixth of capital, less two shares and one-fourth, equal to six shares. Reduce this, by removing the shares which are to be subtracted, and adding

---

$$1 - x - y = 6v$$

$$x = v + \tfrac{1}{5}\left[\tfrac{1}{4} - v\right]; \; y = v - \tfrac{1}{4}\left[\tfrac{1}{3} - x - v\right]$$

i.e. $\tfrac{2}{3} + \tfrac{1}{3} - x - v + \tfrac{1}{4}\left[\tfrac{1}{3} - x - v\right] = 6v$

or $\tfrac{2}{3} + \tfrac{5}{4}\left[\tfrac{1}{3} - x - v\right] = 6v$

or $\tfrac{2}{3} + \tfrac{5}{4}\left[\tfrac{1}{3} - \tfrac{1}{4} + \tfrac{1}{4} - v - \tfrac{1}{5}\left[\tfrac{1}{4} - v\right] - v\right] = 6v$

or $\tfrac{2}{3} + \tfrac{5}{4}\left[\tfrac{1}{12} + \tfrac{4}{5}\left[\tfrac{1}{4} - v\right] - v\right] = 6v$

$\therefore \tfrac{2}{3} + \tfrac{5}{4 \times 12} + \tfrac{1}{4} = \left[7 + \tfrac{5}{4}\right]v$

$\therefore \tfrac{8}{3} + \tfrac{5}{12} + 1 = 33v$ $\therefore \tfrac{49}{12 \times 33} = \tfrac{49}{396} = v$

$\therefore x = v + \tfrac{10}{396}$, and $y = v - \tfrac{6}{396}$

them to the other shares. Then you have one and one-eighth of one-sixth of capital, equal to eight shares and one-fourth. Reduce this to one capital, by subtracting from the parts as much as one forty-ninth of them. Then you have a capital equal to eight shares and four forty-ninths. Assume now every share to be forty-nine; then the entire capital will be three hundred and ninety-six; the share forty-nine; the legacy

(86) determined by one-fourth, ten; and the exception from the second share will be six.

On the Legacy with a Dirhem.

" A man dies, and leaves four sons, and bequeaths to some one a dirhem, and as much as the share of a son, and one-fourth of what remains from one-third after the deduction of that share." Computation :* Take

------

* Let the capital $= 1$; a dirhem $= \delta$;

the legacy $= x$; and a son's share $= v$

$$1 - x = 4v$$

$$x = v + \tfrac{1}{4} \left[\tfrac{1}{3} - v \right] + \delta$$

$$\therefore \tfrac{2}{3} + \tfrac{1}{3} - v - \tfrac{1}{4} \left[\tfrac{1}{3} - v \right] - \delta = 4v$$

$$\therefore \tfrac{2}{3} + \tfrac{3}{4} \left[\tfrac{1}{3} - v \right] - \delta = 4v$$

$$\therefore \tfrac{2}{3} + \tfrac{1}{4} - \delta = \left[4 + \tfrac{3}{4} \right] v$$

$$\therefore \tfrac{11}{12} - \delta = \tfrac{19}{4} v$$

$\therefore \tfrac{11}{57}$ of the capital $- \tfrac{12}{57}$ of a dirhem $= v$

and $\tfrac{13}{57}$ of the capital $+ \tfrac{48}{57}$ of a dirhem $= x$, the legacy.

If we assume the capital to be so many dirhems, or a dirhem to be such a part of the capital, we shall obtain the one-third of the capital and subtract from it one share; there remains one-third, less one share. Then subtract one-fourth of the remainder, namely, one-fourth of one-third, less one-fourth of the share; then subtract also one dirhem; there remain three-fourths of one-third of the capital, that is, one-fourth of the capital, less three-fourths of the share, and less one dirhem. Add this to two-thirds of the capital. The sum is eleven-twelfths of the capital, less three-fourths of the share and less one dirhem, equal to four shares. Reduce this by removing three-fourths of the share and one dirhem; then you have eleven-twelfths of the capital, equal to four shares and three-fourths, plus one dirhem. Complete your capital, by adding to the shares and one dirhem one-eleventh of the same. Then you have the capital equal to five shares and two-elevenths and one dirhem and one-eleventh. If you (87) wish to exhibit the dirhem distinctly, do not complete your capital, but subtract one from the eleven on account of the dirhem, and divide the remaining ten by the portions, which are four and three-fourths. The quotient is two and two-nineteenths of a dirhem. Assuming, then, the capital to be twelve dirhems, each

---

value of the son's share in terms of a dirhem, or of the capital only.

Thus, if we assume the capital to be 12 dirhems,

$$v = \tfrac{12}{57}\,[11-1]\,\delta = \tfrac{120}{57}\,\delta = 2\tfrac{2}{19}\ \text{dirhems},$$

$$x = \tfrac{12}{57}\,[13+4]\,\delta = \tfrac{204}{57}\,\delta = 3\tfrac{11}{19}\ \text{dirhems}.$$

share will be two dirhems and two-nineteenths. Or, if you wish to exhibit the share distinctly, complete your square, and reduce it, when the dirhem will be eleven of the capital.

If he leaves five sons, and bequeaths to some person a dirhem, and as much as the share of one of the sons, and one-third of what remains from one-third, and again, one-fourth of what remains from the one-third after the deduction of this, and one dirhem more; then the computation is this:* You take one-third, and subtract one share; there remains one-third less one share. Subtract herefrom that which is still in your hands, namely, one-third of one-third less one-third of the share. Then subtract also the dirhem; there remain two-thirds of one-third, less two-thirds of the share and less one dirhem. Then subtract one-fourth of what you have, that is, one-eighteenth, less one-sixth of a share and less one-fourth of a dirhem, and

---

* Let the legacy $=x$; and a son's share $=v$

$$1-x=5v$$

$$\tfrac{2}{3}+\tfrac{1}{3}-v-\tfrac{1}{3}\left[\tfrac{1}{3}-v\right]-\eth-\tfrac{1}{4}\left[\tfrac{2}{3}\left[\tfrac{1}{3}-v\right]-\eth\right]-\eth=5v$$

i.e. $\tfrac{2}{3}+\tfrac{2}{3}\left[\tfrac{1}{3}-v\right]-\eth-\tfrac{1}{4}\left[\tfrac{2}{3}\left[\tfrac{1}{3}-v\right]-\eth\right]-\eth=5v$

i.e. $\tfrac{2}{3}+\tfrac{3}{4}\left[\tfrac{2}{3}\left[\tfrac{1}{3}-v\right]-\eth\right]-\eth=5v$

$\therefore \tfrac{2}{3}+\tfrac{1}{6}-\tfrac{1}{2}v-\tfrac{7}{4}\eth=5v$

$\therefore \tfrac{5}{6}-\tfrac{7}{4}\eth=\tfrac{11}{2}v$

$\therefore \tfrac{10}{66}$ of the capital $-\tfrac{21}{66}$ of a dirhem $=v$

$\therefore \tfrac{16}{66}$ of the capital $+\tfrac{105}{66}$ of a dirhem $=x$, the legacy.

If the capital $=\tfrac{45}{2}$ dirhems, or $\tfrac{1}{3}$ of the capital $=7\tfrac{1}{2}$ dirhems,

$v=\tfrac{34}{11}$ dirhems $=3\tfrac{1}{11}$ dirhems.

subtract also the second dirhem; the remainder is half
one-third of the capital, less half a share and less one
dirhem and three-fourths; add thereto two-thirds of the
capital, the sum is five-sixths of the capital, less one
half of a share, and less one dirhem and three-fourths,
equal to five shares. Reduce this, by removing the (88)
half share and the one dirhem and three-fourths,
and adding them to the (five) shares. Then you
have five-sixths of capital, equal to five shares and a
half plus one dirhem and three-fourths. Complete
your capital, by adding to five shares and a half and
to one dirhem and three-fourths, as much as one-fifth
of the same. Then you have the capital equal to six
shares and three-fifths plus two dirhems and one-
tenth. Assume, now, each share to consist of ten
parts, and one dirhem likewise of ten; then the capital is eighty-seven parts. Or, if you wish to exhibit
the dirhem distinctly, take the one-third, and subtract
from it the share; there remains one-third, less one
share. Assume the one-third (of the capital) to be
seven and a half (dirhems). Subtract one-third of what
you have, namely, one-third of one-third;* there
remain two-thirds of one-third, less two-thirds of the
share: that is, five dirhems, less two-thirds of the
share. Then subtract one, on account of the one
dirhem, and you retain four dirhems, less two-thirds

* There is an omission here of the words " less one third
of a share."

of the share. Subtract now one-fourth of what you have, namely, one part less one-sixth of a share; and remove also one part on account of the one dirhem; the remainder, then, is two parts less half a share. Add this to the two-thirds of the capital, which is fifteen (dirhems). Then you have seventeen parts less half a share, equal to five shares. Reduce this, by removing half a share, and adding it to the five shares. Then it is seventeen parts, equal to (89) five shares and a half. Divide now seventeen by five and a half; the quotient is the value of one share, namely, three dirhems and one-eleventh; and one-third (of the capital) is seven and a half (dirhems).

If he leaves four sons, and bequeaths to some person as much as the share of one of his sons, less one-fourth of what remains from one-third after the deduction of the share, and one dirhem; and to another one-third of what remains from the one-third, and one dirhem; then this legacy is determined by one-third.*

---

* Let the 1st legacy be x, the 2d y; and a son's share $= v$

$$1 - x - y = 4v$$

i.e. $\frac{2}{3} + \frac{1}{3} - v + \frac{1}{4}[\frac{1}{3} - v] - \eth - \frac{1}{3}[\frac{1}{3} - v + \frac{1}{4}(\frac{1}{3} - v) - \eth] - \eth = 4v$

i.e. $\frac{2}{3} + \frac{2}{3}[\frac{1}{3} - v + \frac{1}{4}(\frac{1}{3} - v) - \eth] - \eth = 4v$

i.e. $\frac{2}{3} + \frac{2}{3}[\frac{5}{4}(\frac{1}{3} - v) - \eth] - \eth = 4v$

$\therefore \frac{2}{3} + \frac{5}{18} - \frac{5}{6}v - \frac{5}{3}\eth = 4v$

$\therefore \frac{17}{18} - \frac{5}{3}\eth = \frac{29}{6}v$

$\therefore \frac{17}{87} - \frac{20}{58}\eth = v$

also $\frac{14}{87} + \frac{33}{58}\eth = x$

$\frac{5}{87} + \frac{47}{58}\eth = y$

Take one-third of the capital, and subtract from it one share; there remains one-third, less one share; add hereto one-fourth of what you have: then it is one-third and one-fourth of one-third, less one share and one-fourth. Subtract one dirhem; there remains one-third of one and one-fourth, less one dirhem, and less one share and one-fourth. There remains from the one-third as much as five-eighteenths of the capital, less two-thirds of a dirhem, and less five-sixths of a share. Now subtract the second dirhem, and you retain five-eighteenths of the capital, less one dirhem and two-thirds, and less five-sixths of a share. Add to this two-thirds of the capital, and you have seventeen-eighteenths of the capital, less one dirhem and two-thirds, and less five-sixths of a share, equal to four shares. Reduce this, by removing the quantities which are to be subtracted, and adding them to the shares; then you have seventeen-eighteenths of the capital, equal to four portions and five-sixths plus one dirhem and two-thirds. Complete your capital by (90) adding to the four shares and five-sixths, and one dirhem and two-thirds, as much as one-seventeenth of the same. Assume, then, each share to be seventeen, and also one dirhem to be seventeen.* The whole capital will then be one hundred and seventeen. If you wish to exhibit the dirhem distinctly, proceed with it as I have shown you.

---

* Capital $= \frac{8\,7}{1\,7} v + \frac{3\,0}{1\,7} \eth$ ∴ if $v = 17$, and $\eth = 17$, capital $= 117$

If he leaves three sons and two daughters, and bequeaths to some person as much as the share of a daughter plus one dirhem; and to another one-fifth of what remains from one-fourth after the deduction of the first legacy, plus one dirhem; and to a third person one-fourth of what remains from one-third after the deduction of all this, plus one dirhem; and to a fourth person one-eighth of the whole capital, requiring all the legacies to be paid off by the heirs generally: then you calculate this by exhibiting the dirhems distinctly, which is better in such a case.* Take one-fourth of the capital, and assume it to be six dirhems; the entire capital will be twenty-four dirhems. Subtract one share from the one-fourth; there remain six dirhems less one share. Subtract also one dirhem; there remain five dirhems less one share. Subtract

---

* Let the legacies to the three first legatees be, severally, x, y, z; the fourth legacy $= \frac{1}{8}$; and let a daughters' share $= v$.

$$\therefore \tfrac{7}{8} - x - y - z = 8v$$

$$x = v + \delta; \quad y = \tfrac{1}{5}\left[\tfrac{1}{4} - x\right] + \delta; \quad z = \tfrac{1}{4}\left[\tfrac{1}{3} - x - y\right] + \delta$$

Then $\tfrac{7}{8} - \tfrac{1}{3} + \tfrac{1}{3} - x - y - \tfrac{1}{4}\left[\tfrac{1}{3} - x - y\right] - \delta = 8v$

$$\therefore \tfrac{13}{24} + \tfrac{3}{4}\left[\tfrac{1}{3} - x - y\right] - \delta = 8v$$

but $\tfrac{1}{3} - x - y = \tfrac{1}{3} - \tfrac{1}{4} + \tfrac{1}{4} - x - \tfrac{1}{5}\left[\tfrac{1}{4} - x\right] - \delta$

$$= \tfrac{1}{12} + \tfrac{4}{5}\left[\tfrac{1}{4} - x\right] - \delta$$

$$= \tfrac{1}{12} + \tfrac{1}{5} - \tfrac{4}{5}v - \tfrac{9}{5}\delta$$

$$= \tfrac{17}{60} - \tfrac{4}{5}v - \tfrac{9}{5}\delta$$

$$\therefore \tfrac{13}{24} + \tfrac{3}{4} \times \tfrac{17}{60} - \tfrac{3}{5}v - \left[\tfrac{3}{4} \times \tfrac{9}{5} + 1\right]\delta = 8v$$

$\therefore \tfrac{181}{240} - \tfrac{47}{20}\delta = \tfrac{43}{5}v$ $\therefore v = \tfrac{181}{2064} - \tfrac{564}{2064}\delta$, and $1 = \tfrac{2064}{181}v + \tfrac{564}{181}\delta$

$x = \tfrac{181}{2064} + \tfrac{1500}{2064}\delta; \quad y = \tfrac{67}{2064} + \tfrac{1764}{2064}\delta; \quad z = \tfrac{110}{2064} + \tfrac{1248}{2064}\delta$

one-fifth of this remainder; there remain four dirhems, less four-fifths of a share. Now deduct the second dirhem, and you retain three dirhems, less four-fifths of a share. You know, therefore, that the legacy which was determined by one-fourth, is three dirhems, less four-fifths of a share. Return now to the one-third, which is eight, and subtract from it three dirhems, less four-fifths of a share. There remain five (91) dirhems, less four-fifths of a share. Subtract also one-fourth of this and one dirhem, for the legacy; you then retain two dirhems and three-fourths, less three-fifths of a share. Take now one-eighth of the capital, namely, three; after the deduction of one-third, you retain one-fourth of a dirhem, less three-fifths of a share. Return now to the two-thirds, namely, sixteen, and subtract from them one-fourth of a dirhem less three-fifths of a share; there remain of the capital fifteen dirhems and three-fourths, less three-fifths of a share, which are equal to eight shares. Reduce this, by removing three-fifths of a share, and adding them to the shares, which are eight. Then you have fifteen dirhems and three-fourths, equal to eight shares and three-fifths. Make the division: the quotient is one share of the whole capital, which is twenty-four (dirhems). Every daughter receives one dirhem and one-hundred-and-forty-three one-hundred-and-seventy-second parts of a dirhem.*

---

* $v = \frac{181}{2064}$ of the capital $- \frac{564}{2064}$ of a dirhem. If we assume

If you prefer to produce the shares distinctly, take one-fourth of the capital, and subtract from it one share; there remains one-fourth of the capital less one share. Then subtract from this one dirhem: then subtract one-fifth of the remainder of one-fourth, which is one-fifth of one-fourth of the capital, less one-fifth of the share and less one-fifth of a dirhem; and subtract also the second dirhem. There remain four-fifths of the one-fourth less four-fifths of a share, and less one dirhem and four-fifths. The legacies paid out of one fourth amount to twelve two-hundred-and-(92) fortieths of the capital and four-fifths of a share, and one dirhem and four-fifths. Take one-third, which is eighty, and subtract from it twelve, and four-fifths of a share, and one dirhem and four-fifths, and remove one-fourth of what remains, and one dirhem. You retain, then, of the one-third, only fifty-one, less three-fifths of a share, less two dirhems and seven-twentieths. Subtract herefrom one-eighth of the capital, which is thirty, and you retain twenty-one, less three-fifths of a share, and less two dirhems and seven-twentieths, and two-thirds of the capital, being equal to eight shares. Reduce this, by removing that which is to be subtracted, and adding it to the eight shares. Then you have one hundred and eighty-one parts of the

---

the capital to be equal to 24 dirhems

$$v = \frac{181 \times 24 - 564}{2064} \text{ dirhems} = \frac{4344 - 564}{2064} \eth$$
$$= \frac{3780}{2064} \eth = 1\frac{145}{172} \text{ dirhems.}$$

capital, equal to eight shares and three-fifths, plus two dirhems and seven twentieths. Complete your capital, by adding to that which you have fifty-nine one-hundred-and-eighty-one parts. Let, then, a share be three hundred and sixty-two, and a dirhem likewise three hundred and sixty-two.* The whole capital is then five thousand two hundred and fifty-six, and the legacy out of one-fourth† is one thousand two hundred and four, and that out of one-third is four hundred and ninety-nine, and the one-eighth is six hundred and fifty-seven.

On Completement.

" A woman dies and leaves eight daughters, a mo- (93) ther, and her husband, and bequeaths to some person as much as must be added to the share of a daughter to make it equal to one-fifth of the capital; and to another person as much as must be added to the share of the mother to make it equal to one-fourth of

---

* The capital $= \frac{2064}{181}v + \frac{564}{181}\delta$

If we assume $v = 362$, and $\delta = 362$, the capital $= 5256$

Then $x = 724$; $y = 480$; $z = 499$; $\frac{1}{8}$th of capital $= 657$.

† The text ought to stand " the two first legacies are " instead of " the legacy out of one-fourth is."

The first legacy is 724

The second 480

∴ the first + second legacy   =   1204

the capital."* Computation: Determine the parts of
the residue, which in the present instance are thirteen. Take the capital, and subtract from it one-fifth
of the same, less one part, as the share of a daughter: this being the first legacy. Then subtract also
one-fourth, less two parts, as the share of the mother :
this being the second legacy. There remain eleven-
twentieths of the capital, which, when increased by
three parts, are equal to thirteen parts. Remove now
from thirteen parts the three parts on account of the
three parts (on the other side), and you retain eleven-
twentieths of the capital, equal to ten parts. Complete
the capital, by adding to the ten parts as much as nine-
elevenths of the same; then you find the capital equal
to eighteen parts and two-elevenths. Assume now
each part to be eleven; then the whole capital is two
hundred, each part is eleven; the first legacy will be
twenty-nine, and the second twenty-eight.

If the case is the same, and she bequeaths to
some person as much as must be added to the share
(94) of the husband to make it equal to one-third, and to
another person as much as must be added to the share
of the mother to make it equal to one-fourth; and to a

---

* In this case, the mother has $\frac{2}{13}$; and each daughter has $\frac{1}{13}$ of the residue.

$$1-x-y=13v$$
$$\text{i.e. } 1-\tfrac{1}{5}+v-\tfrac{1}{4}+2v=13v$$
$$\therefore \tfrac{11}{20}=10v \quad \therefore v=\tfrac{11}{200}; \quad x=\tfrac{29}{200}; \quad y=\tfrac{28}{200}$$

third as much as must be added to the share of a daughter to make it equal to one-fifth; all these legacies being imposed on the heirs generally: then you divide the residue into thirteen parts.* Take the capital, and subtract from it one-third, less three parts, being the share of the husband; and one-fourth, less two parts, being the share of the mother; and lastly, one-fifth less one part, being the share of a daughter. The remainder is thirteen-sixtieths of the capital, which, when increased by six parts, is equal to thirteen parts. Subtract the six from the thirteen parts: there remain thirteen-sixtieths of the capital, equal to seven parts. Complete your capital by multiplying the seven parts by four and eight-thirteenths, and you have a capital equal to thirty-two parts and four-thirteenths. Assuming then each part to be thirteen, the whole capital is four hundred and twenty.

If the case is the same, and she bequeaths to some person as much as must be added to the share of the mother to make it one-fourth of the capital; and to another as much as must be added to the portion of a daughter, to make it one-fifth of what remains of the capital, after the deduction of the first legacy; then

---

* $1 - [\frac{1}{3} - 3v] - [\frac{1}{4} - 2v] - [\frac{1}{5} - v] = 13v$

i.e. $1 - \frac{1}{3} - \frac{1}{4} - \frac{1}{5} = 7v$

$\therefore \frac{13}{60} = 7v$

$\therefore v = \frac{13}{420}$

you constitute the parts of the residue by taking them out of thirteen.* Take the capital, and subtract from it one-fourth less two parts; and again, subtract one-fifth of what you retain of the capital, less one part; then look how much remains of the capital after the deduction of the parts. This remainder, namely, three-fifths of the capital, when increased by two parts and three-fifths, will be equal to thirteen parts. Subtract two parts and three-fifths from thirteen parts, there remain ten parts and two-fifths, equal to three-fifths of capital. Complete the capital, by adding to the parts which you have, as much as two-thirds of the same. Then you have a capital equal to seventeen parts and one-third. Assume a part to be three, then the capital is fifty-two, each part three; the first legacy will be seven, and the second six.

If the case is the same, and she bequeaths to some person as much as must be added to the share of the mother to make it one-fifth of the capital, and to another one-sixth of the remainder of the capital; then

---

$$* \quad 1-x-y=13v$$
$$x=\tfrac{1}{4}-2v; \quad y=\tfrac{1}{5}[1-x]-v$$
$$1-x-\tfrac{1}{5}[1-x]+v=13v$$
$$\tfrac{4}{5}[1-x]=12v \quad \therefore \quad \tfrac{4}{5}[\tfrac{3}{4}+2v]=12v$$
$$\therefore \tfrac{3}{5}=[12-\tfrac{8}{5}]v=\tfrac{52}{5}v$$
$$\therefore v=\tfrac{3}{52} \quad \therefore \quad x=\tfrac{7}{52}, \quad y=\tfrac{6}{52}$$

the parts are thirteen.*   Take the capital, and subtract from it one-fifth less two parts; and again, subtract one-sixth of the remainder. You retain two-thirds of the capital, which, when increased by one part and two-thirds, are equal to thirteen parts. Subtract the one part and two-thirds from the thirteen parts: there remain two thirds of the capital, equal to eleven parts and one-third. Complete your capital, by adding to the parts as much as their moiety; thus you find the capital equal to seventeen parts. Assume now the capital to be eighty-five, and each part five; then the first legacy is seven, and the second thirteen, and the remaining sixty-five are for the heirs.

If the case is the same, and she bequeaths to some person as much as must be added to the share of the mother, to make it one-third of the capital, less that sum which must be added to make the share of a daughter equal to one-fourth of what remains of the capital after the deduction of the above complement; then the parts are thirteen.†   Take the capital, and (96)

---

* $1-x-y=13v$

$x=\frac{1}{5}-2v; \quad y=\frac{1}{6}[1-x]$

$1-x-\frac{1}{6}[1-x]=13v$

$\therefore \frac{5}{6}[1-x]=13v$

$\therefore \frac{5}{6}[\frac{4}{5}+2v]=13v$

$\therefore \frac{2}{3}+\frac{5}{3}v=13v$

$\therefore \frac{2}{3}=\frac{34}{3}v \quad \therefore v=\frac{1}{17}; \quad x=\frac{7}{85}; \quad y=\frac{13}{85}$

† $1-x+y=13v;$ and $x=\frac{1}{3}-2v; \quad y=\frac{1}{4}[1-x]-v$

$\therefore 1-x+\frac{1}{4}[1-x]-v=13v$

$\therefore \frac{5}{4}[1-x]=14v \quad \therefore \frac{5}{4}[\frac{2}{3}+2v]=14v$

$\therefore \frac{5}{6}=\frac{23}{2}v \quad \therefore v=\frac{5}{69}; \quad x-y=\frac{4}{69}$

s

subtract from it one-third less two parts, and add to the remainder one-fourth (of such remainder) less one part; then you have five-sixths of the capital and one part and a half, equal to thirteen parts. Subtract one part and a half from thirteen parts. There remain eleven parts and a half, equal to five-sixths of the capital. Complete the capital, by adding to the parts as much as one-fifth of them. Thus you find the capital equal to thirteen parts and four-fifths. Assume, now, a part to be five, then the capital is sixty-nine, and the legacy four.

" A man dies, and leaves a son and five daughters, and bequeaths to some person as much as must be added to the share of the son to complete one-fifth and one-sixth, less one-fourth of what remains of one-third after the subtraction of the complement."* Take one-third of the capital, and subtract from it one-fifth and one-sixth of the capital, less two (seventh) parts; so that you retain two parts less four one hundred and twentieths of the capital. Then add it to the exception, which is half a part less one one hundred and

---

* Since there are five daughters and one son, each daughter receives $\frac{1}{7}$, and the son $\frac{2}{7}$ of the residue.

$$1 - x = 7v; \quad \tfrac{1}{5} + \tfrac{1}{6} = \tfrac{11}{30}$$
$$\therefore \tfrac{2}{3} + \tfrac{1}{3} - \tfrac{11}{30} + 2v + \tfrac{1}{4}\left[\tfrac{1}{3} - \tfrac{11}{30} + 2v\right] = 7v$$
$$\therefore \tfrac{2}{3} - \tfrac{1}{30} + 2v + \tfrac{1}{4}\left[\tfrac{-1}{30} + 2v\right] = 7v$$
$$\therefore \tfrac{2}{3} + \tfrac{5}{4}\left[\tfrac{-1}{30} + 2v\right] = 7v$$
$$\therefore \tfrac{4}{6} - \tfrac{1}{24} = \tfrac{9}{2}v$$
$$\therefore \tfrac{5}{8} = \tfrac{9}{2}v \quad \therefore v = \tfrac{5}{36}, \text{ and } x = \tfrac{1}{36}$$

twentieth, and you have two parts and a half less five one hundred and twentieths of capital. Add hereto two-thirds of the capital, and you have seventy-five one hundred and twentieths of the capital and two parts and a half, equal to seven parts. Subtract, now, two parts and a half from seven, and you retain seventy-five one hundred and twentieths, or five-eighths, equal to four parts and a half. Complete your capital, by (97) adding to the parts as much as three-fifths of the same, and you find the capital equal to seven parts and one-fifth part. Let each part be five; the capital is then thirty-six, each portion five, and the legacy one.

If he leaves his mother, his wife, and four sisters, and bequeaths to a person as much as must be added to the shares of the wife and a sister, in order to make them equal to the moiety of the capital, less two-sevenths of the sum which remains from one-third after the deduction of that complement; the Computation is this :* If

---

* From the context it appears, that when the heirs of the residue are a mother, a wife, and 4 sisters, the residue is to be divided into 13 parts, of which the wife and one sister, together, take 5 : therefore the mother and 3 sisters, together, take 8 parts. Each sister, therefore, must take not less than $\frac{1}{13}$, nor more than $\frac{2}{13}$. In the case stated at page 102, a sister was made to inherit as much as a wife; in the present case that is not possible; but the widow must take not less than $\frac{3}{13}$; and each sister not more than $\frac{2}{13}$. Probably, in this case, the mother is supposed to inherit $\frac{2}{13}$; the wife $\frac{3}{13}$; each sister $\frac{2}{13}$.

you take the moiety from one-third, there remains one-sixth. This is the sum excepted. It is the share of the wife and the sister. Let it be five (thirteenth) parts. What remains of the one-third is five parts less one-sixth of the capital. The two-sevenths which he has excepted are two-sevenths of five parts less two-sevenths of one-sixth of the capital. Then you have six parts and three-sevenths, less one-sixth and two-sevenths of one-sixth of the capital. Add hereto two-thirds of the capital; then you have nineteen forty-seconds of the capital and six parts and three-sevenths, equal to thirteen parts. Subtract herefrom the six parts and three-sevenths. There remain nineteen forty-seconds of the capital, equal to six parts and four-sevenths. Complete your capital by adding to it its double and four-nineteenths of it. Then you find the capital equal to fourteen parts, and seventy (98) one hundred and thirty-thirds of a part. Assume one part to be one hundred and thirty-three; then the whole capital is one thousand nine hundred and thirty-

---

$$x+5v=\tfrac{1}{2}; \quad 1-x+\tfrac{2}{7}\left[\tfrac{1}{3}-x\right]=13v$$
$$\therefore \tfrac{2}{3}+\tfrac{1}{3}-x+\tfrac{2}{7}\left[\tfrac{1}{3}-x\right]=13v$$
$$\therefore \tfrac{2}{3}+\tfrac{9}{7}\left[\tfrac{1}{3}-x\right]=13v$$
$$\therefore \tfrac{2}{3}+\tfrac{9}{7}\left[\tfrac{-1}{6}+5v\right]=13v$$
$$\therefore \tfrac{2}{3}-\tfrac{3}{14}=\left[13-\tfrac{45}{7}\right]v$$
$$\therefore \tfrac{19}{42}=\tfrac{46}{7}v \quad \therefore \tfrac{19}{276}=v$$
$$\therefore x=\tfrac{29}{276}, \text{ and the residue}=\tfrac{247}{276}$$

The author unnecessarily takes $7 \times 276 = 1932$ for the common denominator.

two; each part is one hundred and thirty-three, the completion of it is three hundred and one, and the exception of one-third is ninety-eight, so that the remaining legacy is two hundred and three. For the heirs remain one thousand seven hundred and twenty-nine.

COMPUTATION OF RETURNS.[*]

On Marriage in Illness.

" A man, in his last illness, marries a wife, paying (a marriage settlement of) one hundred dirhems, besides which he has no property, her dowry being

---

[*] The solutions which the author has given of the remaining problems of this treatise, are, mathematically considered, for the most part incorrect. It is not that the problems, when once reduced into equations, are incorrectly worked out; but that in reducing them to equations, arbitrary assumptions are made, which are foreign or contradictory to the data first enounced, for the purpose, it should seem, of forcing the solutions to accord with the established rules of inheritance, as expounded by Arabian lawyers.

The object of the lawyers in their interpretations, and of the author in his solutions, seems to have been, to favour heirs and next of kin; by limiting the power of a testator, during illness, to bequeath property, or to emancipate slaves; and by requiring payment of heavy ransom for slaves whom a testator might, during illness, have directed to be emancipated.

ten dirhems. Then the wife dies, bequeathing one-third of her property. After this the husband dies."*
Computation : You take from the one hundred that which belongs entirely to her, on account of the dowry, namely, ten dirhems ; there remain ninety dirhems, out of which she has bequeathed a legacy. Call the sum given to her (by her husband, exclusive of her dowry) thing ; subtracting it, there remain ninety dirhems less thing. Ten dirhems and thing are already in her hands; she has disposed of one-third of her property, which is three dirhems and one-third, and one-third of thing; there remain six dirhems and

---

* Let s be the sum, including the dowry, paid by the man, as a marriage settlement; d the dowry; x the gift to the wife, which she is empowered to bequeath if she pleases.

She may bequeath, if she pleases, $d+x$; she actually does bequath $\frac{1}{3}[d+x]$; the residue is $\frac{2}{3}[d+x]$, of which one half, viz. $\frac{1}{3}[d+x]$ goes to her heirs, and the other half reverts to the husband

∴ the husband's heirs have $s-[d+x]+\frac{1}{3}[d+x]$ or $s-\frac{2}{3}[d+x]$; and since what the wife has disposed of, exclusive of the dowry, is x, twice which sum the husband is to receive, $s-\frac{2}{3}[d+x]=2x$ ∴ $\frac{1}{8}[3s-2d]=x$. But $s=100$; $d=10$ ∴ $x=35$; $d+x=45$; $\frac{1}{3}[d+x]=15$. Therefore the legacy which she bequeaths is 15, her husband receives 15, and her other heirs, 15. The husband's heirs receive $2x=70$.

But had the husband also bequeathed a legacy, then, as we shall see presently, the law would have defeated, in part, the woman's intentions.

two-thirds plus two-thirds of thing, the moiety of which, namely, three dirhems and one-third plus one-third of thing, returns as his portion to the husband.* Thus the heirs of the husband obtain (as his share) ninety-three dirhems and one-third, less two-thirds of thing; and this is twice as much as the sum given to (99) the woman, which was thing, since the woman had power to bequeath one third of all which the husband left;† and twice as much as the gift to her is two things. Remove now the ninety-three and one-third, from two-thirds of thing, and add these to the two things. Then you have ninety-three dirhems and one-third equal to two things and two-thirds. One thing is three-eighths of it, namely, as much as three-eighths of the ninety-three and one-third, that is, thirty-five dirhems.

If the question is the same, with this exception only, that the wife has ten dirhems of debts, and that she bequeaths one-third of her capital; then the Computa-

---

* In other cases, as appears from pages 92 and 93, a husband inherits one-fourth of the residue of his wife's estate, after deducting the legacies which she may have bequeathed. But in this instance he inherits half the residue. If she die in debt, the debt is first to be deducted from her property, at least to the extent of her dowry (see the next problem.)

† When the husband makes a bequest to a stranger, the third is reduced to one-sixth. Vide p. 137.

tion is as follows:* Give to the wife the ten dirhems of her dowry, so that there remain ninety dirhems, out of which she bequeaths a legacy. Call the gift to her thing; there remain ninety less thing. At the disposal of the woman is therefore ten plus thing. From this her debts must be subtracted, which are ten dirhems. She retains then only thing. Of this she bequeaths one-third, namely, one-third of thing: there remains two-thirds of thing. Of this the husband receives by inheritance the moiety, namely, one-third of thing. The heirs of the husband obtain, therefore, ninety dirhems, less two-thirds of thing; and this is twice as much as the gift to her, which was thing; that is, two things. Reduce this, by removing the two-thirds of thing from ninety, and adding them to two things. Then you have ninety dirhems, equal to two things and two-thirds. One thing is three-eighths of this; that is to say, thirty-three dirhems and three-fourths, which is the gift (to the wife).

If he has married her, paying (a marriage settle-

---

* The same things being assumed as in the last example, $s-[d+x]$ remains with the husband; d goes to pay the debts of the wife; and $\frac{x}{3}$ reverts from the wife to the husband.

$$\therefore s-d-\tfrac{2}{3}x=2x \quad \therefore \tfrac{3}{8}[s-d]=x$$
$\therefore$ if $s=100$, and $d=10$, $x=33\frac{3}{4}$; she bequeaths $11\frac{1}{4}$; $11\frac{1}{4}$ reverts to her husband; and her other heirs receive $11\frac{1}{4}$. The husband's heirs receive $2x=67\frac{1}{2}$.

ment of one hundred dirhems, her dowry being ten (100)
dirhems, and he bequeaths to some person one-third of
his property; then the computation is this:* Pay to
the woman her dowry, that is, ten dirhems; there remain ninety dirhems. Herefrom pay the gift to her,
thing; then pay likewise to the legatee who is to
receive one-third, thing: for the one-third is divided

---

* This case is distinguished from that in page 133 by
two circumstances; first, that the woman does not make
any bequest; second, that the husband bequeaths one-third
of his property.

Suppose the husband not to make any bequest. Then,
since the woman had at her disposal $d+x$, but did not make
any bequest, $\frac{1}{2}[d+x]$ reverts to her husband; and the like
amount goes to her other heirs.

$$\therefore s-[d+x]+\tfrac{1}{2}[d+x]=2x \quad \therefore x=\tfrac{1}{3}[2s-d]$$

and since $s=100$, and $d=10$; $x=38$; $d+x=48$;
$\frac{1}{2}[d+x]=24$ reverts to the husband, and the like sum goes
to her other heirs; and $2x=76$, belongs to the husband's
heirs.

Now suppose the husband to bequeath one-third of his
property. The law here interferes with the testator's right
of bequeathing; and provides that whatever sum is at the
disposal of the wife, the same sum shall be at the disposal
of the husband; and that the sum to be retained by the
husband's heirs shall be twice the sum which the husband
and wife together may dispose of.

$$\therefore s-\tfrac{1}{2}[d+x]-x=4x$$

$\therefore \frac{1}{11}[2s-d]=x$; if $s=100$, and $d=10$; $x=\frac{190}{11}=17\frac{3}{11}$;
$d+x=27\frac{3}{11}$; $\frac{1}{2}[d+x]=13\frac{7}{11}$ reverts to the husband, and
the like sum goes to the other heirs of the woman; $17\frac{3}{11}$ is
what the husband bequeaths; and $69\frac{1}{11}=4x$ goes to the
husband's heirs.

T

into two moieties between them, since the wife cannot
take any thing, unless the husband takes the same.
Therefore give, likewise, to the legatee who is to have
one-third, thing. Then return to the heirs of the husband. His inheritance from the woman is five dirhems
and half a thing. There remains for the heirs of the
husband ninety-five less one thing and a half, which
is equal to four things. Reduce this, by removing one
thing and a half, and adding it to the four things.
There remain ninety-five, equal to five things and a
half. Make them all moieties; there will be eleven
moieties; and one thing will be equal to seventeen
dirhems and three-elevenths, and this will be the
legacy.

" A man has married a wife paying (a marriage settlement of) one hundred dirhems, her dowry being ten
dirhems; and she dies before him, leaving ten dirhems,
and bequeathing one-third of her capital; afterwards
the husband dies, leaving one hundred and twenty dirhems, and bequeathing to some person one-third of his
capital." Computation :* Give to the wife her dowry,

---

* Let c be the property which the wife leaves, besides d
the dowry, and x the gift from the husband. She bequeaths
$\frac{1}{3} [c+d+x]$; $\frac{1}{3} [c+d+x]$ goes to her husband; and $\frac{1}{3}$
$[c+d+x]$ to her other heirs. The husband leaves property
s, out of which must be paid the dowry, d; the gift to the
wife, x; and the bequest he makes to the stranger, x; and
his heirs receive from the wife's heirs $\frac{1}{3} [c+d+x]$

namely, ten dirhems; then one hundred and ten dirhems remain for the heirs of the husband. From these the (101) gift to the wife is thing, so that there remain one hundred and ten dirhems less thing; and the heirs of the woman obtain twenty dirhems plus thing. She bequeaths one-third of this, namely, six dirhems and two-thirds, and one-third of thing. The moiety of the residue, namely, six dirhems and two-thirds plus one-third of thing, returns to the heirs of the husband: so that one hundred and sixteen and two-thirds, less two-thirds of thing, come into their hands. He has bequeathed one-third of this, which is thing. There remain, therefore, one hundred and sixteen dirhems and two-thirds less one thing and two-thirds, and this is twice as much as the husband's gift to the wife added to his legacy to the stranger, namely, four things. Reduce this, and you find one hundred and sixteen dirhems and two-thirds, equal to five things and two-thirds. Consequently one thing is equal to

---

$s - d - 2x + \frac{1}{3}[c + d + x] = 4x$, according to the law of inheritance.

$$\therefore \ 3s + c - 2d = 17x, \text{ and } x = \frac{3s + c - 2d}{17}$$

If $s = 120$, $c = 10$, and $d = 10$, $x = \frac{350}{17} = 20\frac{10}{17}$

$c + d + x = 40\frac{10}{17}$; $\frac{1}{3}[c + d + x] = 13\frac{9}{17}$

The wife bequeaths $13\frac{9}{17}$; $13\frac{9}{17}$ go to her husband, and $13\frac{9}{17}$ to her other heirs.

The husband bequeaths to the stranger $20\frac{10}{17}$; he gives the same sum to the wife; and $4x = 82\frac{6}{17}$ go to his heirs.

twenty dirhems and ten-seventeenths; and this is the legacy.

———

On Emancipation in Illness.

"Suppose that a man on his death-bed were to eman-cipate two slaves; the master himself leaving a son and a daughter. Then one of the two slaves dies, leaving a daughter and property to a greater amount than his price.*" You take two-thirds of his price, and what the other slave has to return (in order to complete his (102) ransom). If the slave die before the master, then the son and the daughter of the latter partake of the heri-tage, in such proportion, that the son receives as much as the two daughters together. But if the slave die after the master, then you take two-thirds of his value and what is returned by the other slave, and distribute

———

* From the property of the slave, who dies, is to be de-ducted and paid to the master's heirs, first, two-thirds of the original cost of that slave, and secondly what is wanting to complete the ransom of the other slave. Call the amount of these two sums p; and the property which the slave leaves α.

Next, as to the residue of the slaves' property:

First. If the slave dies before the master, the master's son takes $\frac{1}{2}[\alpha-p]$; the master's daughter $\frac{1}{4}[\alpha-p]$, and the slave's daughter $\frac{1}{4}[\alpha-p]$.

Second. If the slave dies after the master; the master's son is to receive $\frac{2}{3}p$, and the master's daughter $\frac{1}{3}p$; and then the master's son takes $\frac{1}{2}[\alpha-p]$, and the slave's daughter $\frac{1}{2}[\alpha-p]$.

it between the son and the daughter (of the master), in such a manner, that the son receives twice as much as the daughter; and what then remains (from the heritage of the slave) is for the son alone, exclusive of the daughter; for the moiety of the heritage of the slave descends to the daughter of the slave, and the other moiety, according to the law of succession, to the son of the master, and there is nothing for the daughter (of the master).

It is the same, if a man on his death-bed emancipates a slave, besides whom he has no capital, and then the slave dies before his master.

If a man in his illness emancipates a slave, besides whom he possesses nothing, then that slave must ransom himself by two-thirds of his price. If the master has anticipated these two-thirds of his price and has spent them, then, upon the death of the master, the slave must pay two-thirds of what he retains.* But if the master has anticipated from him his whole price and spent it, then there is no claim against the slave, since he has already paid his entire price.

" Suppose that a man on his death-bed emancipates a slave, whose price is three hundred dirhems, not having any property besides; then the slave dies, leaving three hundred dirhems and a daughter." The

---

* The slave retains one-third of his price; and this he must redeem at two-thirds of its value; namely at $\frac{2}{3} \times \frac{1}{3} = \frac{2}{9}$ of his original price.

computation is this :* Call the legacy to the slave thing. He has to return the remainder of his price, after the deduction of the legacy, or three hundred less thing. This ransom, of three hundred less thing, belongs to the master. Now the slave dies, and leaves thing and a (103) daughter. She must receive the moiety of this, namely, one half of thing; and the master receives as much. Therefore the heirs of the master receive three hundred less half a thing, and this is twice as much as the legacy, which is thing, namely, two things. Reduce this by removing half a thing from the three hundred, and adding it to the two things. Then you have three hundred, equal to two things and a half. One thing is, therefore, as much as two-fifths of three hundred,

---

* Let the slave's original cost be a; the property which he dies possessed of, α; what the master bequeaths to the slave, in emancipating him, x. Then the net property which the slave dies possessed of is $\alpha+x-a$. $\frac{1}{2}[\alpha+x-a]$ belongs, by law, to the master; and $\frac{1}{2}[\alpha+x-a]$ to the slave's daughter. The master's heirs, therefore, receive the ransom, $a-x$, and the inheritance, $\frac{1}{2}[\alpha+x-a]$; that is, $\frac{1}{2}[\alpha+a-x]$; and on the same principle as the slave, when emancipated, is allowed to ransom himself at two-thirds of his cost, the law of the case is that 2 are to be taken, where 1 is given.

$$\therefore \frac{1}{2}[\alpha+a-x]=2x \quad \therefore x=\frac{1}{5}[\alpha+a]$$

The daughter's share of the inheritance $=\frac{1}{5}[3\alpha-2a]$

The master's heirs receive.......... $\frac{2}{5}[\alpha+a]$

If, as in the example, $\alpha=a$, $x=\frac{2}{5}a$; the daughter's share $=\frac{1}{5}a$; the heirs of the master receive $\frac{4}{5}a$.

namely, one hundred and twenty. This is the legacy
(to the slave,) and the ransom is one hundred and eighty.

" Some person on his sick-bed has emancipated a
slave, whose price is three hundred dirhems; the slave
then dies, leaving four hundred dirhems and ten dirhems of debt, and two daughters, and bequeathing to
a person one-third of his capital; the master has twenty
dirhems debts." The computation of this case is the
following:* Call the legacy to the slave thing; his ransom is the remainder of his price, namely, three hundred less thing. But the slave, when dying, left four
hundred dirhems; and out of this sum, his ransom,
namely, three hundred less thing, is paid to the

---

* Let the slave's original cost $=a$; the property he dies
possessed of $=\alpha$; the debt he owes $=\epsilon$

He leaves two daughters, and bequeaths to a stranger one-
third of his capital.

The master owes debts to the amount μ; where $a=300$;
$\alpha=400$; $\epsilon=10$; $\mu=20$.

Let what the master gives to the slave, in emancipating
him $=x$.

Slave's ransom $=a-x$; slave's property—slave's ransom $=$
$\alpha+x-a$

Slave's property $-$ ransom $-$ debt $=\alpha+x-a-\epsilon$

Legacy to stranger $=\frac{1}{3}\left[\alpha+x-a-\epsilon\right]$

Residue.........$=\frac{2}{3}\left[\alpha+x-a-\epsilon\right]$

The master, and each daughter, are, by law, severally
entitled to $\frac{1}{3}\times\frac{2}{3}\left[\alpha+x-a-\epsilon\right]$

The master's heirs receive altogether $a-x+\frac{2}{9}\left[\alpha+x-a-\epsilon\right]$
or $\frac{7}{9}\left[a-x\right]+\frac{2}{9}\left[\alpha-\epsilon\right]$, which, on the principle that 2

master, so that one hundred dirhems and thing remain in the hands of the slave's heirs. Herefrom are (first) subtracted the debts, namely, ten dirhems; there remain then ninety dirhems and thing. Of this he has bequeathed one-third, that is, thirty dirhems and one-third of thing; so that there remain for the heirs sixty dirhems and two-thirds of thing. Of this the two daughters receive two-thirds, namely, forty dirhems and four-ninths of thing, and the master

(104) receives twenty dirhems and two-ninths of thing, so that the heirs of the master obtain three hundred and twenty dirhems less seven-ninths of thing. Of this the debts of the master must be deducted, namely, twenty dirhems; there remain then three hundred dirhems less

---

are to be taken for 1 given, ought to be made equal to $2x$.

But the author directs that the equation for determining x be

$$\tfrac{7}{9}[a-x]+\tfrac{2}{9}[a-\epsilon]-\mu=2x$$

$$\therefore \ x=\tfrac{1}{25}[7a+2[a-\epsilon]-9\mu] \qquad\qquad =108$$

Hence the slave receives, the debts which he owes, $\epsilon \ = 10$

$+$ the legacy to the stranger $=\tfrac{1}{25}[9[a-\epsilon]-6a-3\mu]= 66$

$+$ the inheritance of 1st daughter $=\tfrac{1}{25}[6[a-\epsilon]-4a-2\mu]= 44$

$+$ the inheritance of 2d daughter $=\tfrac{1}{25}[6[a-\epsilon]-4a-2\mu]= 44$

$$\overline{\text{Total}=\tfrac{1}{25}[21a+4\epsilon-14a-7\mu]=164}$$

And the master takes $\mu+2x=\tfrac{1}{25}[4a-4\epsilon+14a-7\mu]=236$

Had the slave died possessed of no property whatever, his ransom would have been 200.

His ransom, here stated, exclusive of the sum which the master inherits from him, or $a-x$, $=192$.

seven-ninths of thing; and this sum is twice as much as the legacy of the slave, which was thing; or, it is equal to two things. Reduce this, by removing the seven-ninths of thing, and adding them to two things; there remain three hundred, equal to two things and seven-ninths. One thing is as much as nine twenty-fifths of eight hundred, which is one hundred and eight; and so much is the legacy to the slave.

If, on his sick-bed, he emancipates two slaves, besides whom he has no property, the price of each of them being three hundred dirhems; the master having anti-cipated and spent two-thirds of the price of one of them before he dies;* then only one-third of the price

---

* Were there the first slave only, who has paid off two-thirds of his original cost, the master having spent the money, that slave would have to complete his ransom by paying two-ninths of his original cost, that is $66\frac{2}{3}$ (see page 141).

Were there the second slave only, who has paid off none of his original cost, he would have to ransom himself at two-thirds of his cost; that is by paying 200 (see also page 141).

The master's heirs, in the case described in the text, are entitled to receive the same amount from the two slaves jointly, viz. $266\frac{2}{3}$, as they would be entitled to receive, according to the rule of page 141, from the two slaves, sepa-rately; but the payment of the sum is differently distributed; the slave who has paid two-thirds of his ransom being required to pay one-ninth only of his original cost; and the slave who has paid no ransom, being required to pay two-thirds of his own cost, and one-ninth of the cost of the first slave.

of this slave, who has already paid off a part of his ransom, belongs to the master; and thus the master's capital is the entire price of the one who has paid off nothing of his ransom, and one-third of the price of the other who has paid part of it; the latter is one hundred dirhems; the other three hundred dirhems: one-third of the amount, namely, one hundred and thirty-three dirhems and one third, is divided into two moieties among them; so that each of them receives sixty-six dirhems and two-thirds. The first slave, who has already paid two-thirds of his ransom, pays thirty-three dirhems and one-third; for

(105) sixty-six dirhems and two-thirds out of the hundred belong to himself as a legacy, and what remains of the hundred he must return. The second slave has to return two hundred and thirty-three dirhems and one-third.

"Suppose that a man, in his illness, emancipates two slaves, the price of one of them being three hundred dirhems, and that of the other five hundred dirhems; the one for three hundred dirhems dies, leaving a daughter; then the master dies, leaving a daughter likewise; and the slave leaves property to the amount of four hundred dirhems. With how much must every one ransom himself?"* The computation is this: Call

---

* Let A. be the first slave; his original cost a; the property he dies possessed of α; and let B. be the second slave; and his cost b.

the legacy to the first slave, whose price is three hundred dirhems, thing. His ransom is three hundred dirhems less thing. The legacy to the second slave of a price of five hundred dirhems is one thing and two-thirds, and his ransom five hundred dirhems less one thing and two-thirds (*viz.* his price being one and two-thirds times the price of the first slave, whose ransom was thing, he must pay one thing and two-thirds for

---

Let x be that which the master gives to A. in emancipating him.

A.'s ransom is $a-x$; and his property, minus his ransom, is $a-a+x$.

A.'s daughter receives $\frac{1}{2}[a-a+x]$, and the master's heirs receive $\frac{1}{2}[a-a+x]$

Hence the master receives altogether from A.,
$$a-x+\tfrac{1}{2}[a-a+x]=\tfrac{1}{2}[a+a-x.]$$

B.'s ransom is $b-\dfrac{b}{a}x$

The master's heirs receive from A. and B. together $\frac{1}{2}[a+a+2b]-\frac{1}{2a}[a+2b]x$; and this is to be made equal to twice the amount of the legacies to A. and B., that is,
$$\tfrac{1}{2}[a+a+2b]-\tfrac{1}{2a}[a+2b]x=2\frac{a+b}{a}x$$
$$\therefore\ x=a\frac{a+a+2b}{5a+6b}=\frac{1700}{15}=113\tfrac{1}{3}$$

The master's heirs receive from A., $\dfrac{2a[a+a+b]+3ab}{5a+6b}=293\tfrac{1}{3}$

A.'s daughter receives $[a+b]\dfrac{3a-2}{5a+6}a=800\times\frac{600}{4500}=106\tfrac{2}{3}$

The legacy to B. is $b\dfrac{a+a+2b}{5a+6b}=188\tfrac{8}{9}$; his ransom is $b\dfrac{4a+4b-a}{5a+6b}=311\tfrac{1}{9}$

The master's heirs receive from A. and B. together
$$2[a+b]\frac{a+a+2b}{5a+6b}=604\tfrac{4}{9}.$$

his ransom). Now the slave for three hundred dirhems dies, and leaves four hundred dirhems. Out of this his ransom is paid, namely, three hundred dirhems less thing; and in the hands of his heirs remain one hundred dirhems plus thing: his daughter receives the moiety of this, namely, fifty dirhems and half a thing; and what remains belongs to the heirs of the master, namely, fifty dirhems and half a thing. This is added to the three hundred less thing; the sum is three hundred and fifty less half a thing. Add thereto the ransom of the other, which is five hundred dirhems less one thing and two-thirds; thus, the heirs

(106) of the master have obtained eight hundred and fifty dirhems less two things and one-sixth; and this is twice as much as the two legacies together, which were two things and two-thirds. Reduce this, and you have eight hundred and fifty dirhems, equal to seven things and a half. Make the equation; one thing will be equal to one hundred and thirteen dirhems and one-third. This is the legacy to the slave, whose price is three hundred dirhems. The legacy to the other slave is one and two-thirds times as much, namely, one hundred and eighty-eight dirhems and eight-ninths, and his ransom three hundred and eleven dirhems and one-ninth.

" Suppose that a man in his illness emancipates two slaves, the price of each of whom is three hundred dirhems; then one of them dies, leaving five hundred dirhems and a daughter; the master having left a son."

Computation :* Call the legacy to each of them thing; the ransom of each will be three hundred less thing; then take the inheritance of the deceased slave, which is five hundred dirhems, and subtract his ransom, which is three hundred less thing; the remainder of his inheritance will be two hundred plus thing. Of this, one hundred dirhems and half a thing return to the master by the law of succession, so that now altogether four hundred dirhems less a half thing are in the hands of the master's heirs. Take also the ransom of the other slave, namely, three hundred dirhems less thing; then the heirs of the master obtain seven hundred dir-

---

* The first slave is A.; his cost a; his property α; he leaves a daughter.

The second slave is B.; his cost b.

Then (as in page 147) $\frac{1}{2}[\alpha - a + x]$ goes to the daughter; and $x = a \; \frac{\alpha + a + 2b}{5a + 6b}$

The daughter receives $[a + b] \; \frac{3\alpha - 2a}{5a + 6b}$

The master receives from A. $\frac{2a[a + \alpha + b] + 3\alpha b}{5a + 6b}$
and the master receives from A. and B. together $2[a + b] \; \frac{\alpha + a + 2b}{5a + 6b}$

But if $b = a$ .... $x = \frac{1}{11}[\alpha + 3a] = 127\frac{3}{11}$

The daughter receives .... $\frac{2}{11}[3\alpha - 2a] = 163\frac{7}{11}$

The master receives from A. ...... $\frac{1}{11}[5\alpha + 4a] = 336\frac{4}{11}$

The master receives from B. .... $\frac{1}{11}[8a - \alpha] = 172\frac{8}{11}$

The master receives from A and B. .. $\frac{4}{11}[\alpha + 3a] = 509\frac{1}{11}$

If $b = 0$,

The daughter receives $\frac{1}{5}[3\alpha - 2a]$

The master .... $\frac{2}{5}[\alpha + a]$, as in page 142.

hems less one thing and a half, and this is twice as much as the sum of the two legacies of both, namely (107) two things, consequently as much as four things. Remove from this the one thing and a half: you find seven hundred dirhems, equal to five things and a half. Make the equation. One thing will be one hundred and twenty-seven dirhems and three-elevenths.

" Suppose that a man in his illness emancipate a slave, whose price is three hundred dirhems, but who has already paid off to his master two hundred dirhems, which the latter has spent; then the slave dies before the death of the master, leaving a daughter and three hundred dirhems."* Computation: Take the property left by the slave, namely, the three hundred, and add thereto the two hundred, which the master has spent; this together makes five hundred dirhems. Subtract from this the ransom, which is three hundred less thing

---

* The slave A. dies before his master, and leaves a daughter. His cost is a, of which he has redeemed $\acute{a}$, which the master has spent; and he leaves property α.

Then the daughter receives .. $\frac{1}{2}\left[\alpha+\acute{a}-a+x\right]$

The master receives altogether $\frac{1}{2}\left[\alpha+\acute{a}+a-x\right]$

The master's heirs receive.... $\frac{1}{2}\left[\alpha-\acute{a}+a-x\right]$

And $\frac{1}{2}\left[\alpha-\acute{a}+a-x\right]=2x$ $\therefore$ $x=\frac{1}{5}\left[\alpha-\acute{a}+a\right]$

Hence the daughter receives $\frac{1}{5}\left[3\alpha+2\acute{a}-2a\right]=140$

The master's heirs .... $\frac{1}{5}\left[2\alpha-2\acute{a}+2a\right]=160$

The master receives, in toto, $\frac{1}{5}\left[2\alpha+3\acute{a}+2a\right]=360$

If the slave had not advanced, or the master had not spent $\acute{a}$, the daughter would have received $\frac{1}{5}\left[3\alpha+3\acute{a}-2a\right]=180$ and the master would have received $\frac{1}{5}\left[2\alpha+2\acute{a}+2a\right]=320$.

(since his legacy is thing); there remain two hundred
dirhems plus thing. The daughter receives the moiety
of this, namely, one hundred dirhems plus half a thing;
the other moiety, according to the laws of inheritance,
returns to the heirs of the master, being likewise one
hundred dirhems and half a thing. Of the three hundred dirhems less thing there remain only one hundred
dirhems less thing for the heirs of the master, since
two hundred are spent already. After the deduction
of these two hundred which are spent, there remain
with the heirs two hundred dirhems less half thing, and
this is equal to the legacy of the slave taken twice; or
the moiety of it, one hundred less one-fourth of thing,
is equal to the legacy of the slave, which is thing. Remove from this the one-fourth of thing; then you have
one hundred dirhems, equal to one thing and one-
fourth. One thing is four-fifths of it, namely, eighty
dirhems. This is the legacy; and the ransom is two
hundred and twenty dirhems. Add the inheritance of the
slave, which is three hundred, to two hundred, which (108)
are spent by the master. The sum is five hundred
dirhems. The master has received the ransom of two
hundred and twenty dirhems; and the moiety of the
remaining two hundred and eighty, namely, one hundred and forty, is for the daughter. Take these from
the inheritance of the slave, which is three hundred;
there remain for the heirs one hundred and sixty dirhems, and this is twice as much as the legacy of the
slave, which was thing.

" Suppose that a man in his illness emancipates a slave, whose price is three hundred dirhems, but who has already advanced to the master five hundred dirhems; then the slave dies before the death of his master, and leaves one thousand dirhems and a daughter. The master has two hundred dirhems debts."* Computation: Take the inheritance of the slave, which is one thousand dirhems, and the five hundred, which the master has spent. The ransom from this is three hundred less thing. There remain therefore twelve hundred plus thing. The moiety of this belongs to the daughter : it is six hundred dirhems plus half a thing. Subtract it from the property left by the slave, which was one

---

* A.'s price is a; he has advanced to his master $á$; he leaves property u. He dies before his master, and leaves a daughter.

The master's debts are μ; x is what A. receives, in being emancipated; $a-x$ is the ransom; $\frac{1}{2}[u+á-a+x]$ is what the daughter receives.

Then $u-\frac{1}{2}[u+á-a+x]$ is what remains to the master; and $u-\frac{1}{2}[u+á-a+x]-\mu$ is what remains to him, after paying his debts; and this is to be made equal to $2x$.

Whence $x=\frac{1}{5}[u+a-á-2\mu]$

Hence the daughter receives $\dots\dots\frac{1}{5}[3u-2a+2á-\mu]=640$

The mother receives, inclusive of the debt $\Big\}\dots\dots\dots\frac{1}{5}[2u+2a-2á+\mu]=360$

The master receives, exclusive of the debt $\Big\}\dots\dots\frac{1}{5}[2u+[2a-2á-4\mu]=160$

If the mode given in page 142 had been followed, it would have given $x=\frac{1}{5}[u+a+á-2\mu]$

and the daughter's portion $=\frac{1}{5}[3u-2a+3á-\mu]=740$.

thousand dirhems: there remain four hundred dirhems less half thing. Subtract herefrom the debts of the master, namely, two hundred dirhems; there remain two hundred dirhems less half thing, which are equal to the legacy taken twice, which is thing; or equal to two things. Reduce this, by means of the half thing. Then you have two hundred dirhems, equal to two things and a half. Make the equation. You find one thing, equal to eighty dirhems; this is the legacy. Add now the property left by the slave to the sum which he has (109) advanced to the master: this is fifteen hundred dirhems. Subtract the ransom, which is two hundred and twenty dirhems; there remain twelve hundred and eighty dirhems, of which the daughter receives the moiety, namely, six hundred and forty dirhems. Subtract this from the inheritance of the slave, which is one thousand dirhems: there remain three hundred and sixty dirhems. Subtract from this the debts of the master, namely, two hundred dirhems; there remain then one hundred and sixty dirhems for the heirs of the master, and this is twice as much as the legacy of the slave, which was thing.

" Suppose that a man on his sick-bed emancipates a slave, whose price is five hundred dirhems, but who has already paid off to him six hundred dirhems. The master has spent this sum, and has moreover three hundred dirhems of debts. Now the slave dies, leaving his mother and his master, and property to the amount of seventeen hundred and fifty dirhems, with two hundred

dirhems debts." Computation:* Take the property left
by the slave, namely, seventeen hundred and fifty dirhems, and add to it what he has advanced to the master, namely, six hundred dirhems; the sum is two
thousand three hundred and fifty dirhems. Subtract
from this the debts, which are two hundred dirhems,
and the ransom, which is five hundred dirhems less
thing, since the legacy is thing; there remain then
sixteen hundred and fifty dirhems plus thing. The
mother receives herefrom one-third, namely, five hundred and fifty plus one-third of thing. Subtract now
this and the debts, which are two hundred dirhems,
from the actual inheritance of the slave, which is
seventeen hundred and fifty; there remain one thou-
(110) sand dirhems less one-third of thing. Subtract from
this the debts of the master, namely, three hundred

---

* A. dies before his master, and leaves a mother. His
price was a; he has redeemed $\acute{a}$, which the master has
spent. The property he leaves is α. He owes debts ς.
The master owes debts μ.

$\frac{1}{3}[\alpha + \acute{a} - a + x - \varsigma]$ is the mother's.

$\alpha - \frac{1}{3}[\alpha + \acute{a} - a + x - \varsigma] - \varsigma$ is the master's.

$\alpha - \frac{1}{3}[\alpha + \acute{a} - a + x - \varsigma] - \varsigma - \mu = 2x =$ the master's, after
paying his debts.

Hence .... $x = \frac{1}{7}[2\alpha + a - \acute{a} - 2\varsigma - 3\mu] = 300$

Mother's................ $= \frac{1}{7}[3\alpha - 2a + 2\acute{a} - 3\varsigma - \mu] = 650$

Master's, without μ .... $= \frac{1}{7}[4\alpha + 2a - 2\acute{a} - 4\varsigma - 6\mu] = 600$

Mother's, with μ .... $= \frac{1}{7}[4\alpha + 2a - 2\acute{a} - 4\varsigma + \mu] = 900$

A. receives, inclusive of ς $= \frac{1}{7}[3\alpha - 2a + 2\acute{a} + 4\varsigma - \mu] = 850$.

dirhems; there remain seven hundred dirhems less one-third of thing. This is twice as much as the legacy of the slave, which is thing. Take the moiety: then three hundred and fifty less one-sixth of thing are equal to one thing. Reduce this, by means of the one-sixth of thing; then you have three hundred and fifty, equal to one thing and one-sixth. One thing will then be equal to six-sevenths of the three hundred and fifty, namely, three hundred dirhems; this is the legacy. Add now the property left by the slave to what the master has spent already; the sum is two thousand three hundred and fifty dirhems. Subtract herefrom the debts, namely, two hundred dirhems, and subtract also the ransom, which is as much as the price of the slave less the legacy, that is, two hundred dirhems; there remain nineteen hundred and fifty dirhems. The mother receives one-third of this, namely, six hundred and fifty dirhems. Subtract this and the debts, which are two hundred dirhems, from the property actually left by the slave, which was seventeen hundred and fifty dirhems; there remain nine hundred dirhems. Subtract from this the debts of the master, which are three hundred dirhems; there remain six hundred dirhems, which is twice as much as the legacy.

" Suppose that some one in his illness emancipates a slave, whose price is three hundred dirhems: then the slave dies, leaving a daughter and three hundred dirhems; then the daughter dies, leaving her husband and

three hundred dirhems; then the master dies." Computation:* Take the property left by the slave, which is
three hundred dirhems, and subtract the ransom, which
(111) is three hundred less thing; there remains thing, one
half of which belongs to the daughter, while the other
half returns to the master. Add the portion of the
daughter, which is half one thing, to her inheritance,
which is three hundred; the sum is three hundred dirhems plus half a thing. The husband receives the moiety
of this; the other moiety returns to the master, namely
one hundred and fifty dirhems plus one-fourth of thing.
All that the master has received is therefore four hundred and fifty less one-fourth of thing; and this is twice
as much as the legacy; or the moiety of it is as much as

---

* A. is emancipated by his master, and then dies, leaving
a daughter, who dies, leaving a husband. Then the master
dies.

A.'s price $= a$; his property α. What he receives from
the master $= x$.

The daughter's property $= \delta$

A.'s ransom $= a - x$. The daughter inherits $\frac{1}{2}[\alpha - a + x]$,
and $\frac{1}{2}[\alpha - a + x]$ goes to the master.

$\qquad \frac{1}{2}[\delta + \frac{1}{2}[\alpha - a + x]]$ goes to the daughter's husband

$\qquad \qquad$ and $\frac{1}{2}[\delta + \frac{1}{2}[\alpha - a + x]]$ to the master.

Hence, according to the author, we are to make

$$a - x + \frac{1}{2}[\alpha - a + x] + \frac{1}{2}[\delta + \frac{1}{2}[\alpha - a + x]] = 2x$$

$$\therefore \quad x = \frac{1}{9}[3\alpha + a + 2\delta] = 200$$

Daughter's share $= \frac{1}{9}[6\alpha - 4a + \delta] = 100$

Husband's $\ldots \ldots = \frac{1}{9}[3\alpha - 2a + 5\delta] = 200$

Master's $\ldots \ldots \ldots = \frac{1}{9}[2\alpha + 6a + 4\delta] = 400.$

the legacy itself, namely, two hundred and twenty-five dirhems less one-eighth thing are equal to thing. Reduce this by means of one-eighth of thing, which you add to thing; then you have two hundred and twenty-five dirhems, equal to one thing and one-eighth. Make the equation: one thing is as much as eight-ninths of two hundred and twenty-five, namely, two hundred dirhems.

" Suppose that some one in his illness emancipates a slave, of the price of three hundred dirhems; the slave dies, leaving five hundred dirhems and a daughter, and bequeathing one-third of his property; then the daughter dies, leaving her mother, and bequeathing one-third of her property, and leaving three hundred dirhems." Computation:* Subtract from the property left

---

* A. is emancipated, and dies, leaving a daughter, and bequeathing one-third of his property to a stranger.

The daughter dies, leaving a mother, and bequeathing one-third of her property to a stranger.

A.'s price is a ; his property is α

The daughter's property is δ.

A.'s ransom is $a-x$; $\alpha-a+x$ is his property, clear of ransom.

$\frac{1}{3}[\alpha-a+x]$ goes to the stranger; and the like amount to A.'s daughter, and to the master.

$\frac{1}{3}[3\delta+\alpha-a+x]$ is the property left by the daughter.

$\frac{1}{9}[3\delta+\alpha-a+x]$ is the bequest of the daughter to a stranger.

$\frac{2}{9}[3\delta+\alpha-a+x]$ is the residue, of which $\frac{1}{3}$d,

viz. $\frac{2}{27}[3\delta+\alpha-a+x]$ is the mother's,

and $\frac{4}{27}[3\delta+\alpha-a+x]$ is the master's;

by the slave his ransom, which is three hundred dirhems less thing; there remain two hundred dirhems
plus thing. He has bequeathed one-third of his property, that is, sixty-six dirhems and two-thirds plus
one-third of thing. According to the law of succession,
(112) sixty-six dirhems and two-thirds and one-third of thing
belong to the master, and as much to the daughter.
Add this to the property left by her, which is three
hundred dirhems: the sum is three hundred and sixty-
six dirhems and two-thirds and one-third of thing.
She has bequeathed one-third of her property, that is,
one hundred and twenty-two dirhems and two-ninths
and one-ninth of thing; and there remain two hundred
and forty-four dirhems and four-ninths and two-ninths
of thing. The mother receives one-third of this,
namely, eighty-one dirhems and four-ninths and one-
third of one-ninth of a dirhem plus two-thirds of one-
ninth of thing. The remainder returns to the master;
it is a hundred and sixty-two dirhems and eight-ninths
and two-thirds of one-ninth of a dirhem plus one-ninth
and one-third of one-ninth of thing, as his share of the
heritage.

Hence, according to the author, we are to make

$$a-x+\tfrac{1}{3}\left[\alpha-a+x\right]+\tfrac{4}{27}\left[3\delta+\alpha-a+x\right]=2x$$

Therefore . . . . $x=\tfrac{1}{68}\left[13\alpha+14a+12\delta\right]=210\tfrac{5}{17}$

The daughter's share . . $=\tfrac{1}{68}\left[27\alpha-18a+4\delta\right]=136\tfrac{4}{17}$

The daughter's bequest $=\tfrac{1}{68}\left[9\alpha-6a+24\delta\right]=145\tfrac{10}{17}$

The mother's share . . . . $=\tfrac{2}{68}\left[3\alpha-2a+8\delta\right]=\ 97\tfrac{1}{17}$

The master's . . . . $=\tfrac{2}{68}\left[13\alpha+14a+12\delta\right]=420\tfrac{10}{17}$.

Thus the master's heirs have obtained five hundred and twenty-nine dirhems and seventeen twenty-sevenths of a dirhem less four-ninths and one-third of one-ninth of thing; and this is twice as much as the legacy, which is thing. Halve it: You have two hundred and sixty-four dirhems and twenty-two twenty-sevenths of a dirhem, less seven twenty-sevenths of thing. Reduce it by (113) means of the seven twenty-sevenths which you add to the one thing. This gives one hundred and sixty-four dirhems and twenty-two twenty-sevenths, equal to one thing and seven twenty-sevenths of thing. Make the equation, and adjust it to one single thing, by subtracting from it as much as seven thirty-fourths of the same. Then one thing is equal to two hundred and ten dirhems and five-seventeenths; and this is the legacy.

" Suppose that a man in his illness emancipates a slave, whose price is one hundred dirhems, and makes to some one a present of a slave-girl, whose price is five hundred dirhems, her dowry being one hundred dirhems, and the receiver cohabits with her." Abu Hanifah says: The emancipation is the more important act, and must first be attended to.

Computation :* Take the price of the girl, which is

---

* The price of the slave-girl being a ; and what she receives on being emancipated x, her ransom is $a-x$.

If her dowry is a, he that receives her, takes $a+x$.

five hundred dirhems ; and remember that the price of the slave is one hundred dirhems. Call the legacy of the donee thing. The emancipation of the slave, whose price is one hundred dirhems, has already taken place. He has bequeathed one thing to the donee. Add the dowry, which is one hundred dirhems less one-fifth thing. Then in the hands of the heirs are six hundred dirhems less one thing and one-fifth of thing. This is twice as much as one hundred dirhems and thing; the moiety of it is equal to the legacy of the two, namely, three hundred less three-fifths of thing. Reduce this by removing the three-fifths of thing from three hundred, and add the same to one thing. This gives three hundred dirhems, equal to one thing and three-fifths and one hundred dirhems. Subtract now from three hun-

---

Hence, according to the author, we are to make

$$a - x = 2 [\alpha + x]; \quad \text{whence } x = \frac{a - 2\alpha}{3}$$

And her ransom is $\frac{2}{3} [a + \alpha]$

But if a male slave be at the same time emancipated by the master, the donee must pay the ransom of that slave. If his price was b, $b - \frac{b}{a} x$ is his ransom.

Hence, according to the author, we are to make the sum of the two ransoms, viz. $a - x + b - \frac{b}{a} x = 2 [\alpha + x]$

$$\therefore a + b - 2\alpha = [3 + \frac{b}{a}] x \quad \therefore x = a \frac{a + b - 2\alpha}{3a + b} = 125$$

The donee pays ransom, in respect of the slave-girl $(a - x) = 375$

and he pays ransom for the male slave $\ldots\ldots\ldots b - \frac{b}{a} x = 75$.

dred the one hundred, on account of the other one hundred. There remain two hundred dirhems, equal to one thing and three-fifths. Make the equation with this. One thing will be five-eighths of what you have ; (114) take therefore five-eighths of two hundred. It is one hundred and twenty-five. This is thing; it is the legacy to the person to whom he had presented the girl.

" Suppose that a man emancipates a slave of a price of one hundred dirhems, and makes to some person a present of a slave girl of the price of five hundred dirhems, her dowry being one hundred dirhems; the donee cohabits with her, and the donor bequeaths to some other person one-third of his property." According to the decision of Abu Hanifah, no more than one-third can be taken from the first owner of the slave-girl; and this one-third is to be divided into two equal parts between the legatee and the donee. Computation:* Take the price of the girl, which is five hundred dirhems. The legacy out of this is thing; so that the heirs obtain five hundred dirhems less thing; and the dowry is one hundred less one-fifth of thing; consequently they

---

* The same notation being used as in the last example, the equation for determining x, according to the author, is to be

$$a - x + b - \frac{b}{a}x - x = 2\left[a + 2x\right]$$

$$\therefore \; x = \frac{a}{6a+b}\left[a+b-2a\right] = 64\tfrac{16}{31}.$$

Y

obtain six hundred dirhems less one thing and one-fifth of thing. He bequeaths to some person one third of his capital, which is as much as the legacy of the person who has received the girl, namely, thing. Consequently there remain for the heirs six hundred less two things and one-fifth, and this is twice as much as both their legacies taken together, namely, the price of the slave plus the two things bequeathed as legacies. Halve it, and it will by itself be equal to these legacies: it is then three hundred less one and one-tenth of thing. Reduce this by means of the one and one-tenth of thing. Then you have three hundred, equal to three things and one-tenth, plus one hundred dirhems. Remove one hundred on occount of (the opposite) one hundred; there remain two hundred, equal to three things and one-tenth. Make now the reduction. One thing will be as much as thirty-one (115) parts of the sum of dirhems which you have; and just so much will be the legacy out of the two hundred; it is sixty-four dirhems and sixteen thirty-one parts.

" Suppose that some one emancipates a slave girl of the price of one hundred dirhems, and makes to some person a present of a slave girl, which is five hundred dirhems worth; the receiver cohabits with her, and her dowry is one hundred dirhems; the donor bequeaths to some other person as much as one-fourth of his capital." Abu Hanifah says: The master of the girl cannot be required to give up more than one-third, and the legatee, who is to receive one-fourth, must give up one-fourth. Computation :* The price of the girl is five hundred dirhems. The legacy out of this is thing; there remain five hundred dirhems less thing. The dowry is one hundred dirhems less one-fifth of thing; thus the heirs obtain six hundred dirhems less one and one-fifth of thing. Subtract now the legacy of the person to whom one-fourth has been bequeathed, namely, three-fourths of thing; for if one-third is thing then one-fourth is as much as three-fourths of the same.

There remain then six hundred dirhems less one thing and thirty-eight fortieths. This is equal to the legacy taken twice. The moiety of it is equal to the legacies by themselves, namely, three hundred dirhems less thirty-nine fortieths of thing. Reduce this by means of the latter fraction. Then you have three hun- (116) dred dirhems, equal to one hundred dirhems and two things and twenty-nine fortieths. Remove one hundred on account of the other one hundred. There remain two hundred dirhems, equal to two things and twenty-nine-fortieths. Make the equation. You will then find one thing to be equal to seventy-three dirhems and forty-three one-hundred-and-ninths dirhems.

---

* The same notation being used as in the two former examples, the equation for determining x, according to the author, is

$$a - x + b - \frac{b}{a} \ x - \tfrac{3}{4}x = 2\left[\alpha + 1\tfrac{3}{4}x\right]$$

Whence $x = \dfrac{4a}{21a + 4b}\left[a + b - 2a\right] = 73\tfrac{43}{109}.$

On return of the Dowry.

" A MAN, in the illness before his death, makes to
some one a present of a slave girl, besides whom he
has no property. Then he dies. The slave girl is
worth three hundred dirhems, and her dowry is one
hundred dirhems. The man to whom she has been
presented, cohabits with her." Computation:* Call
the legacy of the person to whom the girl is presented, thing. Subtract this from the donation: there
remain three hundred less thing. One-third of this
difference returns to the donor on account of dowry
(since the dowry is one-third of the price): this is
one hundred dirhems less one-third of thing. The
donor's heirs obtain, therefore, four hundred less
one and one-third of thing, which is equal to twice
the legacy, which is thing, or to two things. Transpose the one and one-third thing from the four hundred, and add it to the two things; then you have four
hundred, equal to three things and one-third. One
thing is, therefore, equal to three-tenths of it, or to one
hundred and twenty dirhems, and this is the legacy.

---

* Let a be the slave-girl's price — α her dowry.
Then, according to the author, we are to make

$$a - x + \alpha - \frac{\alpha}{a}\, x = 2x$$

Therefore $x = \frac{a}{3a + \alpha}\, [a + \alpha] = \frac{3}{10} \times 400 = 120$

The donee is to receive the girl's dowry, worth 400, for 280.

" Or, suppose that he, in his illness, has made a present of the slave girl, her price being three hundred, her dowry one hundred dirhems; and the donor dies, after having cohabited with her." Computation :* Call the legacy thing: the remainder is three hundred less thing. The donor having cohabited with her, the dowry remains with him, which is one-third of the legacy, since the dowry is one-third of the price, or one-third of thing. Thus the donor's heirs obtain three (117) hundred less one and one-third of thing, and this is twice as much as the legacy, which is thing, or equal to two things. Remove the one and one-third of thing, and add the same to the two things. Then you have three hundred, equal to three things and one-third. One thing is, therefore, three-tenths of it, namely ninety dirhems. This is the legacy.

If the case be the same, and both the donor and donee have cohabited with her; then the Computation

---

* If the donor has cohabited with the slave-girl, the donor's heirs are to retain the dowry, but must allow the donee, in addition to the legacy x, the further sum of $\frac{\alpha}{a} x$;

The ransom is then $a - x - \frac{\alpha}{a} x$, which according to the author is to be made equal to $2x$.

$$\text{Whence } x = \frac{a^2}{3x + a} = 90$$

The donee is to receive the girl, worth 300, for 210.

is this :* Call the legacy thing; the deduction is three hundred dirhems less thing. The donor has ceded the dowry to the donee by (the donee's) having cohabited with her: this amounts to one-third of thing: and the donee cedes one-third of the deduction, which is one hundred less one-third of thing. Thus, the donor's heirs obtain four hundred less one and two-thirds of thing, which is twice as much as the legacy. Reduce this, by separating the one and two-thirds of thing from four hundred, and add them to the two things. Then you have four hundred things, equal to three things and two-thirds. One thing of these is three-elevenths of four hundred; namely, one hundred and

---

* If the donor has previously cohabited with the slave-girl, it appears from the last example, that the donee is entitled to ransom her for $a - x - \frac{a}{a}x$.

If the donee cohabits with the slave-girl, it appears from the last example but one, that he is entitled to redeem the dowry, α, for $\alpha - \frac{a}{a}x$

The redemption of the girl and dowry is

$$a - x - \frac{a}{a}x + \alpha - \frac{a}{a}x,$$

which, according to the author, is to be made equal to $2x$.

That is $a + \alpha - \frac{a + 2\alpha}{a}x = 2x$

Whence $x = \frac{a}{3a + 2\alpha} \times [a + \alpha] = 109\frac{1}{11}$

The donee is to receive the girl and dowry, worth 400, for $290\frac{10}{11}$.

nine dirhems and one-eleventh. This is the legacy. The deduction is one hundred and ninety dirhems and ten-elevenths. According to Abu Hanifah, you call the thing a legacy, and what is obtained on account of the dowry is likewise a legacy.

If the case be the same, but that the donor, having cohabited with her, has bequeathed one-third of his (118) capital, then Abu Hanifah says, that the one-third is halved between the donee and the legatee. Computation:* Call the legacy of the person to whom the slave-girl has been given, thing. After the deduction of it, there remain three hundred, less thing. Then take the dowry, which is one-third of thing; so that the donor retains three hundred less one and one-third of thing: the donee's legacy being, according to Abu Hanifah, one and one-third of thing; according to other lawyers, only thing. The legatee, to whom one-third is bequeathed, receives as much as the legacy of the donee, namely, one and one-third of thing. The donor thus retains three hundred, less two things and

---

* The second case is here solved in a different way.

$$a-x-\frac{\alpha}{a}x = 2\left[x+\frac{a}{a}x\right]$$

$$\therefore x=\frac{a^2}{3[\alpha+a]}$$

This being halved between the legatee and donee becomes

$$\frac{a^2}{6[\alpha+a]}=37\tfrac{1}{2}$$

The donee receives the girl, worth 300, for $262\tfrac{1}{2}$.

two-thirds – equal to twice the two legacies, which are two things and two-thirds. The moiety of this, namely, one hundred and fifty less one and one-third of thing, must, therefore, be equal to the two legacies. Reduce it, by removing one and one-third of thing, and adding the same to the two legacies (things). Then you find one hundred and fifty, equal to four things. One thing is one-fourth of this, namely, thirty-seven and a half.

If the case be, that both the receiver and the donor have cohabited with her, and the latter has disposed of one-third of his capital by way of legacy; then the computation,* according to Abu Hanifah, is, that you call the legacy thing. After the deduction of it, there remain three hundred less thing. Then the dowry is taken, which is one hundred less one-third of thing; so that there are four hundred dirhems less one and one-third of thing. The sum returned from the dowry is one-third of thing; and the legatee, who is to receive one-third, obtains as much as the legacy of the first, namely, thing and one-third of thing. Thus there

---

* According to the author's rule, which is purely arbitrary,

$$a - 2x + a - \frac{3a}{a}x = 4\left[1 + \frac{a}{a}\right]x$$

Whence $x = a\ \dfrac{a+a}{6a+7a} = 48$

The donee will have to redeem the girl and dowry, worth 400, for 352.

remain four hundred dirhems less three things, equal
to twice the legacy, namely, two things and two-thirds. (119)
Reduce this, by means of the three things, and you find
four hundred, equal to eight things and one-third.
Make the equation with this: one thing will be forty-
eight dirhems.

" Suppose that a man on his sick-bed makes to another a present of a slave-girl, worth three hundred dirhems, her dowry being one hundred dirhems; the
donee cohabits with her, and afterwards, being also on
his sick-bed, makes a present of her to the donor,
and the latter cohabits with her. How much does he
acquire by her, and how much is deducted?"* Com-

---

* We have here the only instance in the treatise of a
simple equation, involving two unknown quantities. For
what the donee receives is one unknown quantity; and what
the donor receives back again from the donee, called by the
author " part of thing," is the other unknown quantity.

Let what the donee receives $=x$, and what the donor
receives $=y$.

Then, retaining the same notation as before, according to
the author, the donee receives, on the whole

$$x-y-[\alpha-\tfrac{\alpha}{a}x]+\tfrac{\alpha}{a}[x-y]=2y$$

and the donor receives, on the whole

$$a-x+y+[\alpha-\tfrac{\alpha}{a}x]-\tfrac{\alpha}{a}[x-y]=2[x+\tfrac{\alpha}{a}[x-y]]$$

Whence $x=\tfrac{1}{2}\dfrac{a}{4a^2+5a\alpha-\alpha^2}[3a^2+3a\alpha-2\alpha^2]=102$

$$y=\tfrac{1}{2}\dfrac{a}{4a^2+5a\alpha-\alpha^2}[a^2-2\alpha^2]=21$$

z   But

putation : Take the price, which is three hundred dirhems; the legacy from this is thing; there remain with
the donor's heirs three hundred less thing; and the
donee obtains thing. Now the donee gives to the
donor part of thing : consequently, there remains only
thing less part of thing for the donee. He returns to the
donor one hundred less one-third of thing ; but takes the
dowry, which is one-third of thing, less one-third of
part of thing. Thus he obtains one and two-thirds
thing less one hundred dirhems and less one and one-
third of part of thing. This is twice as much as part
of thing; and the moiety of it is as much as part of
thing, namely, five-sixths of thing less fifty dirhems
and less two-thirds of part of thing. Reduce this by
removing two-thirds of part of thing and fifty dirhems.
Then you have five-sixths of thing, equal to one and
two-thirds of part of thing plus fifty dirhems. Reduce
this to one single part of thing, in order to know what
the amount of it is. You effect this by taking three-fifths
(120) of what you have. Then one part of thing plus thirty
dirhems is equal to half a thing; and one-half thing
less thirty dirhems is equal to part of thing, which is
the legacy returning from the donee to the donor.
Keep this in memory.

Then return to what has remained with the donor;

But the reasons for reducing the question to these two
equations are not given by the author, and seem to depend
on the dicta of the sages of the Arabian law.

this was three hundred less thing : hereto is now added the part of thing, or one-half thing less thirty dirhems. Thus he obtains two hundred and seventy less half one thing. He further takes the dowry, which is one hundred dirhems less one-third thing, but has to return a dowry, which is one-third of what remains of thing after the subtraction of part of thing, namely, one-sixth of thing and ten dirhems. Thus he retains three hundred and sixty less thing, which is twice as much as thing and the dowry, which he has returned. Halve it : then one hundred and eighty less one-half thing are equal to thing and that dowry. Reduce this, by removing one-half thing and adding it to the thing and the dowry : you find one hundred and eighty dirhems, equal to one thing and a half plus the dowry which he has returned, and which is one-sixth thing and ten dirhems. Remove these ten dirhems; there remain one hundred and seventy dirhems, equal to one and two-thirds things. Reduce this, in order to ascertain what the amount of one thing is, by taking three-fifths of what you have; you find that one hundred and two are equal to thing, which is the legacy from the donor to the donee : and the legacy from the donee to the donor is the moiety of this, less thirty dirhems, namely, twenty-one.

On Surrender in Illness.

(121) " Suppose that a man, on his sick-bed, deliver to some one thirty dirhems in a measure of victuals, worth ten dirhems; he afterwards dies in his illness; then the receiver returns the measure and returns besides ten dirhems to the heirs of the deceased." Computation: He returns the measure, the value of which is ten dirhems, and places to the account of the deceased twenty dirhems; and the legacy out of the sum so placed is thing; thus the heirs obtain twenty less thing, and the measure. All this together is thirty dirhems less thing, equal to two things, or equal to twice the legacy. Reduce it by separating the thing from the thirty, and adding it to the two things. Then, thirty are equal to three things. Consequently, one thing must be one-third of it, namely, ten, and this is the sum which he obtains out of what he places to the account of the deceased.

" Suppose that some one on his sick-bed delivers to a person twenty dirhems in a measure worth fifty dirhems; he then repeals it while still on his sick bed, and dies after this. The receiver must, in this case, return four-ninths of the measure, and eleven dirhems and one-ninth."* Computation : You know that the

---

* Let a be the gift of money; and the value of the measure $m \times a$.

It appears from the context that the donee is to pay the heirs $\frac{2}{3}ma$.

price of the measure is two and a half times as much as
the sum which the donor has given the donee in money;
and whenever the donee returns anything from the
money capital, he returns from the measure as much
as two and a half times that amount. Take now from
the measure as much as corresponds to one thing, that
is, two things and a half, and add this to what remains
from the twenty, namely, twenty less thing. Thus the
heirs of the deceased obtain twenty dirhems and one (122)
thing and a half. The moiety of this is the legacy,
namely, ten dirhems and three-fourths of thing; and
this is one-third of the capital, namely, sixteen dirhems
and two-thirds. Remove now ten dirhems on account
of the opposite ten; there remain six dirhems and two-
thirds, equal to three-fourths of thing. Complete the
thing, by adding to it as much as one-third of the
same; and add to the six dirhems and two-thirds

It is arbitrary how he shall apportion this sum between the
money capital and the measure.

If he pays on the money capital $p.\,a$
and on the measure $\dots\dots\dots q.\,ma$
we have the equation $p.\,a + q.\,ma = \frac{2}{3}\,ma$

or $p\ \ +q\ m\ =\frac{2}{3}\,m$

The author assumes $p = \frac{m}{2}.\,q$

Whence $q = \frac{4}{9}$, and $p = \frac{5}{9}$, and therefore the donee pays
on the money capital$\dots\dots$ $\frac{5}{9}\,a = 11\frac{1}{9}$
and on the measure $\dots\dots\frac{4}{9}\,ma = 22\frac{2}{9}$

Total $\dots\dots\dots\dots\dots 33\frac{1}{3}.$

likewise one-third of the same, namely, two dirhems and two-ninths; this yields eight dirhems and eight-ninths, equal to thing. Observe now how much the eight dirhems and eight-ninths are of the money capital, which is twenty dirhems. You will find them to be four-ninths of the same. Take now four-ninths of the measure and also five-ninths of twenty. The value of four-ninths of the measure is twenty-two dirhems and two-ninths; and the five-ninths of the twenty are eleven dirhems and one-ninth. Thus the heirs obtain thirty-three dirhems and one-third, which is as much as two-thirds of the fifty dirhems.—God is the Most Wise!

NOTES.

———

Page 1, *line* 2-5

The neglected state of the manuscript, in which most diacritical points are wanting, makes me very doubtful whether I have correctly understood the author's meaning in several passages of his preface.

In the introductory lines, I have considered the words التي باداء ما افترض منها علي من يعبده من خلقه as an amplification of what might briefly have been expressed by التي باداءها " through the performance of which." I conceive the author to mean, that God has prescribed to man certain duties, ان الله قد افترض علي الناس شيئا من (باداء ما افترض &c.), المحامد, and that by performing these (&c. المحامد,) we express our thankfulness (نقع اسم الشكر) &c.

Since my translation was made, I have had the advantage of consulting Mr. SHAKESPEAR about this passage. He prefers to read تستوجب, تقع and تومن instead of نقع, نستوجب, and نومن, and proposes to translate as follows: " Praise to God for his favours in that which is proper for him from among his laudable deeds, which in the performance of what he has rendered indispensible from (or by reason of) them on (the part of) whoever of his creatures worships him, gives the name of thanksgiving, and secures the increase, and preserves from deterioration."

The construction here assumed is evidently easier than that adopted by myself, in as far as the relative pronoun التي representing محامده, is made the subject of the three subsequent verbs تقع, &c., whilst my translation presumes a transition from the third person (as in ما هو أهله, and in من الله) to the first (as in ونقع يعبده, &c.).

A marginal note in the manuscript explains the words "The لعل تقديره ونومن صاحبه من الغير by ونومن من الغير meaning may be : we preserve from change him who enjoys it," (viz. the divine bounty, taking صاحبه for صاحب نعم الله. The *change* here spoken of is the forfeiture of the divine mercy by bad actions; for "God does not *change* the mercy which he bestows on men, as long as they do not *change* that which is within themselves." بان الله لم يك مغيرا نعمة انعمها علي قوم حتي يغيروا ما بانفسهم (Coran, Sur. VIII. v. 55, ed. Hinck.).

Page 1, line 7.

علي حين فترة من الرسل] See *Coran*, Sur. v. v. 22. ed. *Hinck*.

Page 1, line 14, 15.

I am particularly doubtful whether I have correctly read and translated the words of the text from واحتسابا to وذكره. Instead of احتسابا للاجر I should have preferred احسانا

للاخر " benefitting others," if the verb احسن could be construed with the preposition ل .

Page 2, line 1.

To the words رجل سبق a marginal note is given in the
manuscript, which is too much mutilated to be here tran-
scribed, but which mentions the names of several authors
who first wrote on certain branches of science, and con-
cludes with asserting, that the author of the present treatise
was the first that ever composed a book on Algebra.

Page 2, line 4.

An interlinear note in the manuscript explains فلم سعته
by جمع مفترقه .

Page 2, line 10.

MOHAMMED gives no definition of the science which he
intends to treat of, nor does he explain the words جبر jebr,
and مقابلة mokābalah, by which he designates certain ope-
rations peculiar to the solution of equations, and which,
combined, he repeatedly employs as an expression for this
entire branch of mathematics. As the former of these words
has, under various shapes, been introduced into the several
languages of Europe, and is now universally used as the
designation of an important division of mathematical science,
I shall here subjoin a few remarks on its original sense, and
on its use in Arabic mathematical works.

The verb جبر jabar of which the substantive جبر jebr is
derived, properly signifies to restore something broken,

especially to cure a fractured bone. It is thus used in the following passage from MOTANABBI (p. 143, 144, *ed. Calcutt.*)

يا من الوذ به فيما اوملـه   ومن اعوذ به بمـا احـادره
ومن توهمت ان البـحر راخته   جودا وان عطاياد جواهـره
ارحم شباب فتي اودت بجدته   يد البلا وذوي في السجن ناضره
لايجبر الناس عظما انت كاسره   ولا يهيضون عظما انت جابـره

"O thou on whom I rely in whatever I hope, with whom I seek refuge from all that I dread; whose bounteous hand seems to me like the sea, as thy gifts are like its pearls : pity the youthfulness of one, whose prime has been wasted by the hand of adversity, and whose bloom has been stifled in the prison. Men will not heal a bone which thou hast broken, nor will they break one which thou hast healed."

Hence the Spanish and Portuguese expression *algebrista* for a person who heals fractures, or sets right a dislocated limb.

In mathematical language, the verb جبر means, to make perfect, or to complete any quantity that is incomplete or liable to a diminution; *i. e.* when applied to equations, to transpose negative quantities to the opposite side by changing their signs. The negative quantity thus removed is construed with the particle ب : thus, if $x^2 - 6 = 23$ shall be changed into $x^2 = 29$, the direction is اجبر المال بالستة وزدها علي الثلثة والعشرين *i. e.* literally "Restore the square from (the deficiency occasioned to it by) the six, and add these to the twenty-three."

The verb جبر is not likewise used, when in an equation an integer is substituted for a fractional power. of the unknown quantity: the proper expression for this is either the second or fourth conjugation of كمل, or the second of تمّ.

The word مقابلة *mokābalah* is a noun of action of the verb قبل to be in front of a thing, which in the third conjugation is used in a reciprocal sense of two objects being opposite one another or standing face to face; and in the transitive sense of putting two things face to face, of confronting or comparing two things with one another.

In mathematical language it is employed to express the comparison between positive and negative terms in a compound quantity, and the reduction subsequent to such comparison. Thus $100 + 10x - 10x + x^2$ is reduced to $100 + x^2$ بعد ان قابلنا به " after we have made a comparison."

When applied to equations, it signifies, to take away such quantities as are the same and equal on both sides. Thus the direction for reducing $x^2 + x = x^2 + 4$ to $x = 4$ will be expressed by قابل.

In either application the verb requires the preposition ب before a pronoun implying the entire equation or compound quantity, within which the comparison and subsequent reduction is to take place.

The verb قابل is not likewise used, when the reduction of an equation is to be performed by means of a division: the proper term for this operation being كسر.

The mathematical application of the substantives جبر and مقابلة will appear from the following extracts.

1. A marginal note on one of the first leaves of the Oxford manuscript lays down the following distinction :

اما الجبر وهو اتمام كل شيء ناقص بما يتم من غير جنسه والمقابلة من المفاعلة وهو المواجهة ولهذا يقال للمصلي القبلة اذا واجهها فلما صار لهذا الحساب جزيل عمله جبر الناقص [بما] نقص منه وزيادة مثل ما جبر به الناقص علي الجنس المقابل لتقابل الزيادة مثلما جبر به الناقص وكثر الاستعمال في ذلك فسمي جبرا ومقابلة لانه يجبر كل شيء بما نقص منه و تقابل الاجناس بعضها الي بعض .... وقد صارت المقابلة ايضا تعرف [عند] اهل الحساب حذف المقادير المتشابهة

" *Jebr* is the restoration of anything defective by means of what is complete of another kind. *Mokābalah*, a noun of action of the third conjugation, is the facing a thing : whence it is applied to one praying, who turns his face towards the *kiblah*. In this branch of calculation, the method commonly employed is the restoring of something defective in its deficiency, and the adding of an amount equal to this restoration to the other side, so as to make the completion (on the one side) and this addition (on the other side) to face (or to balance) one another. As this method is frequently resorted to, it has been named *jebr* and *mokābalah* (or Restoring and Balancing), since here every thing is made complete if it is deficient, and the opposite sides are made to balance one another. . . . . . . Mathematicians also take the word *mokábalah* in the sense of the removal of equal quantities (from both sides of an equation)."

According to the first part of this gloss, in reducing $x - 5a = 10a$ to $x = 15a$, the substitution of x in place of $x - 5a$ would afford an instance of *jebr* or restoration, and the corresponding addition of $5a$ to $10a$, would be an example of *mokábalah* or balancing. From the following extracts it will be seen, that *mokábalah* is more generally taken in the sense stated last by the gloss.

2. HAJI KHALFA, in his bibliographical work (MS. of the British Museum, fol. 167, *recto**.) gives the following explanation : ومعنى الجبر زيادة قدر ما نقص في الجملة المعادلة بالاستثناء في الجملة الاخرى ليتعادلا ومعنى المقابلة اسقاط الزايد من احدى الجملتين للتعادل "*Jebr* is the adding to one side what is negative on the other side of an equation, owing to a subtraction, so as to equalize them. *Mokábalah* is the removal of what is positive from either sum, so as to make them equal."

A little farther on HAJI KHALFA gives further illustration of this by an example : كما في قولنا عشرة الا شيئا يعدل اربعة اشياء فالجبر رفع الاستثناء بان يزاد مثل المستثني علي المستثني منه فيجعل العشرة كاملة كانه يجبر نقصانه ويزاد مثل المستثني علي عديله كزيادة الشيء في المثال بعد جبر العشرة علي اربعة اشياء حتي تصير خمسة فالمقابلة ان تنقص

---

* This manuscript is apparently only an abridgement of HAJI KHALFA's work.

الإجناس من الطرفين بعدّة واحدة قيل هي تقابل بعض
الاشياء ببعض علي المساوات كما في مثال المذكور اذا قوبلت
العشرة بالخمسة علي المساوات وسمي العلم بهذين العلمين
علم الجبر والمقابلة لكثرة وقوعها فيه ...... " For instance if
we say : 'Ten less one thing equal to four things;' then
jebr is the removal of the subtraction, which is performed
by adding to the minuend an amount equal to the subtrahend : hereby the ten are made complete, that which
was defective in them being restored. An amount equal
to the subtrahend is then added to the other side of the
equation : as in the above instance, after the ten have been
made complete, one thing must be added to the four things,
which thus become five things. *Mokābalah* consists in
withdrawing the same amount from quantities of the same
kind on both sides of the equation; or as others say, it is
the balancing of certain things against others, so as to
equalize them. Thus, in the above example, the ten are
balanced against the five with a view to equalize them.
This science has therefore been called by the name of
these two rules, namely, the rule of *jebr* or restoration,
and of *mokābalah* or reduction, on account of the frequent use that is made of them."

3. The following is an extract from a treatise by ABU
ABDALLAH AL-HOSAIN BEN AHMED,* entitled, المقدمة

---

* I have not been able to find any information about this writer. The
copy of the work to which I refer is comprized in the same volume with
MOHAMMED BEN MUSA's work in the Bodleian library. It bears no date.

الكافية في اصول الجبر و المقابلة or " A complete introduction
to the elements of algebra."

باب تفسير الجبر والمقابلة ۞ اعلم ان الحساب انما سموا
هذا النوع جبرا لانهم وضعوه علي معادلة .... فلما كانوا
وضعوه علي المعادلة اداهم العمل في اكثر مسائلة الي معادلة
الناقص بغير الناقص فلم يكن بد من جبر ذلك الناقص بما
ينقص وزيادة مثل ذلك علي ما عدله فلما كثر ذلك فيه
سموه جبرا فهذا معني الجبر وعلة تسميتهم به هذا النوع ۞
فاما المقابلة فهو حذف المقادير المتشابهة من الجهتين ۞

" On the original meaning of the words *jebr* and
mokābalah. This species of calculation is called *jebr*
(or completion) because the question is first brought to
an equation ...... And as, after the equation has been
formed, the practice leads in most instances to equalize
something defective with what is not defective, that
defective quantity must be completed where it is defective; and an addition of the same amount must be made
to what is equalized to it. As this operation is frequently
employed (in this kind of calculation), it has been called
jebr: such is the original meaning of this word, and
such the reason why it has been applied to this kind of
calculation. *Mokābalah* is the removal of equal magni-
tudes on both sides (of the equation)."

4. In the *Kholāset al Hisāb*, a compendium of arithmetic and geometry by Baha-eddin Mohammed ben al
Hosain (died A.H. 1031, i. e. 1575 A.D.) the Arabic

text of which, together with a Persian commentary by
ROSHAN ALI, was printed at Calcutta* (1812. 8vo.) the
following explanation is given (pp. 334. 335.) والطرف
ون الاستثناء يكمل ويزاد مثل ذلك علي الاخر وهو الجبر
والاجناس المتجانسة المستوية في الطرفين تسقط منهما وهو المقابلة
" The side (of the equation) on which something is to be
subtracted, is made complete, and as much is added to
the other side : this is *jebr* ; again those cognate quantities which are equal on both sides are removed, and this
is *mokābalah*." The examples which soon follow, and
the solution of which BAHA-EDDIN shows at full length,
afford ample illustration of these definitions.   In page 338,
$1500 - \frac{1}{4}x = x$ is reduced to $1500 = 1\frac{1}{4}x$; this he says is
effected by *jebr*. In page 341, $7x = \frac{1}{2}x^2 + \frac{1}{2}x$ is reduced
to $13x = x^2$, and this he states to be the result of both
jebr and *mokābalah*.

The Persians have borrowed the words *jebr* and *mokā-*
balah, together with the greater part of their mathematical terminology, from the Arabs.   The following extract
from a short treatise on Algebra in Persian verse, by
MOHAMMED NADJM-EDDIN KHAN, appended to the Calcutta edition of the *Kholāset al Hisāb*, will serve as an
illustration of this remark.

---

* A full account of this work by Mr. STRACHEY will be found in the
twelfth volume of the Asiatic Researches, and in HUTTON's Tracts on
mathematical and philosophical subjects, vol. II. pp. 179-193.   See also
HUTTON's Mathematical Dictionary, art. *Algebra*.

طرفي كه دروست حرف الآ

تكميل كن و مثل آن را

بر طرف دگر فزون كن اي حبر

در مصطلح است نام اين جبر

هنگام معادله تو بشناس

افتد اگر اين كه بعض اجناس

با وصف تجانس از سويت

در هر طرف اند بي مزيت

بايد كه زهر دو سو براني

نامش تو مقابله بخواني

"Complete the side in which the expression *illā* (less, minus) occurs, and add as much to the other side, O learned man: this is in correct language called *jebr*. In making the equation mark this: it may happen that some terms are cognate and equal on each side, without distinction; these you must on both sides remove, and this you call *mokābalah*."

With the knowledge of Algebra, its Arabic name was introduced into Europe. LEONARDO BONACCI of Pisa, when beginning to treat of it in the third part of his treatise of arithmetic, says: *Incipit pars tertia de solutione quarundam quæstionum secundum modum Algebræ et Almucabalæ, scilicet oppositionis et restaurationis*. That the sense of the Arabic terms is here given in the inverted order, has been remarked by COSSALI. The definitions of *jebr* and *mokābalah* given by another early Italian

2 B

writer, LUCAS PACIOLUS, or LUCAS DE BURGO, are thus reported by COSSALI: *Il commune oggetto dell' operar loro è recare la equazione alla sua maggior unità. Gli uffizj loro per questo commune intento sono contrarj: quello dell'* Algebra *è di restorare li extremi dei diminuti; e quello di* Almucabala *di levare da li extremi i superflui. Intende Fra Luca per* extremi *i membri dell' equazione.*

Since the commencement of the sixteenth century, the word *mokābalah* does no longer appear in the title of Algebraic works. HIERONYMUS CARDAN's Latin treatise, first published in 1545, is inscribed: *Artis magnæ sive de regulis algebraicis liber unus.* A work by JOHN SCHEU-BELIUS, printed at Paris in 1552, is entitled: *Algebræ compendiosa facilisque descriptio, qua depromuntur magna Arithmetices miracula.* (See HUTTON's Tracts, &c. II. pp. 241-243.) PELLETIER's Algebra appeared at Paris in 1558, under the title: *De occulta parte numerorum quam Algebram vocant, libri duo.* (HUTTON, l. c. p. 245. MONTUCLA, *hist. des math.* I. p. 613.) A Portuguese treatise, by PEDRO NUÑEZ or NONIUS, printed at Amberez in 1567, is entitled: *Libro de Algebra y Arithmetica y Geometria.* (MONTUCLA, l. c. p. 615.)

In FEIZI's Persian translation of the *Lilavati* (written in 1587, printed for the first time at Calcutta in 1827, 8vo.) I do not recollect ever to have met with the word جبر ; but مقابله is several times used in the same sense as in the above Persian extract.

Page 3, line 3, seqq.

In the formation of the numerals, the thousand is not, like the ten and the hundred, multiplied by the units only, but likewise by any number of a higher order, such as tens and hundreds: there being no special words in Arabic (as is the case in Sanscrit) for ten-thousand, hundred-thousand, &c.

From this passage, and another on page 10, it would appear that our author uses the word عقد, *plur.* عقود, knot or tie, as a general expression for *all* numerals of a higher order than that of the units. Baron S. DE SACY, in his Arabic Grammar, (vol. I. § 741) when explaining the terms of Arabic grammar relative to numerals, translates عقود by *nœuds*, and remarks: *Ce sont les noms des dixaines, depuis* vingt *jusqu'à* quatre-vingt-dix.

Page 3, line 9-11.

The forms of algebraic expression employed by LEONARDO are thus reported by COSSALI (*Origine, &c. dell' Algebra*, I. p. 1.): *Tre considerazioni distingue Leonardo nel numero : una assoluta, o semplice, ed è quella del numero in se stesso ; le altre due relative, e sono quelle di radice e di quadrato. Nominando il quadrato soggiugne* QUI VIDELICET CENSUS DICITUR, *ed il nome di censo è quello di cui in seguito si serve.* That LEONARDO seems to have chosen the expression *census* on account of its acceptation, which is correspondent to that of the

Arabic مال, has already been remarked by Mr. COLE-
BROOKE (Algebra, &c., Dissertation, p. liv.)

PACIOLO, who wrote in Italian, used the words *numero*,
cosa, and *censo;* and this notation was retained by TAR-
TAGLIA. From the term *cosa* for the unknown number,
exactly corresponding in its acceptation to the Arabic شيء
thing, are derived the expressions *Ars cossica* and the
German *die Coss*, both ancient names of the science of
Algebra. CARDAN's Latin terminology is *numerus*, *quadratum*, and *res*, for the latter also *positio* or *quantitas
ignota*.

Page 3, line 17.

I have added from conjecture the words وحذور تعدل عدد
which are not in the manuscript. There occur several
instances of such omissions in the work.

The order in which our author treats of the simple
equations is, 1st. $x^2 = px$; 2d. $x^2 = n$; 3d. $px = n$. LEO-
NARDO had them in the same order. (See COSSALI, l. c.
p. 2.) In the *Kholáset al Hisáb* the arrangement is, 1st.
$n = px$; 2d. $px = x^2$; 3d. $n = x^2$.

Page 5, line 9.

In the *Lilavati*, the rule for the solution of the case
$cx^2 + bx = a$ is expressed in the following stanza.

गुणघ्नमूलोनयुतस्य राशे
दृष्टस्य युक्तस्य गुणार्धकृत्या १

मूलं गुणार्धेन युतं विहीनं
वर्गीकृतं प्रष्टुरभीष्टराशिः ॥

i. e. rendered literally into Latin:

Per multiplicatam radicem diminutæ [vel] auctæ quantitatis

Manifestæ, additæ ad dimidiati multiplicatoris quadratum

Radix, dimidiato multiplicatore addito [vel] subtracto,

In quadratum ducta—est interrogantis desiderata

quantitas.

The same is afterwards explained in prose: यो
राशिः स्वमूलेन केनचित् गुणितेन द्विनो युतो
वा दृष्टस्तस्य मूलस्य गुणार्धकृत्या युक्तस्य
दृष्टस्य यत् पदं तद्गुणार्धेन युतं यदि
मूलोनो दृष्टो राशिर्भवति यदि गुणघ्नमूलयुतो
दृष्टस्तर्हि विहीनं कार्यं तस्य वर्गो राशिः
स्यात् ॥ i. e. "A quantity, increased or diminished
by its square-root multiplied by some number, is given.
Then add the square of half the multiplier of the root to
the given number: and extract the square-root of the
sum. Add half the multiplier, if the difference were
given; or subtract it, if the sum were so. The square of
the result will be the quantity sought." (Mr. COLEBROOKE's
translation.)

FEIZI's Persian translation of this passage runs thus:

هرگاه شخصي عدديرا مضمر کرد وجذر اورا یا کسري

از جذر اورا در عددی ضرب کرد ونام مضروب فیه بیان کرد
وحاصل ضرب را با عدد مضمر جمع کرد یا ازوی نقصان کرد
آنچه بعد از جمع یا نقصان حاصل شده است آنرا نیز
ظاهر کرد طریق دانستن آن عدد چنان است که مضروب
فیه مذکوررا تنصیف کرده مجذور او بگیرند وبا حاصل جمع
یا باقی نقصان که ظاهر کرده بود جمع کرده جذرش بگیرند
بعد از آن نصف مضروب فیه مذکوررا با جذر مذکور جمع
کنند اگر سائل نقصان کرده باشد ونقصان کنند اگر او جمع
کرده است بعد ازآن مجمع یا باقی را مجذور بگیرند بعینه
همان عدد مضمر خواهد بود ✡

With the above Sanskrit stanza from the *Lilavati* some
readers will perhaps be interested to compare the following
Latin verses, which MONTUCLA (I. p. 590) quotes from
LUCAS PACIOLUS:

Si res et census numero coœquantur, a rebus
Dimidio sumpto, censum producere debes,
Addereque numero, cujus a radice totiens
Tolle semis rerum, census latusque redibit.

Page 6, *line* 16.

فنصف الاجذار تکون خمسة] Such instances of the
common instead of the apocopate future, after the impe-
rative, are too frequent in this work, than that they could
be ascribed to a mere mistake of the copyist: I have
accordingly given them as I found them in the manuscript.

Page 7, *line* 1.

وكذلك فافعل [The same structure occurs page 21, line 15.

Page 8, *line* 11.

[فيذه الستة الضروب Hadji Khalfa, in his article on Algebra, quotes the following observation from Ibn Khal-

DUN. قال ابن خلدون وقد بلغنا ان بعض أئمة التعاليم من اهل المشرق انتهى المعادلات الي اكثر من هذه الستة وبلغها الي فوق العشرين واستخرج لها كلها اعمالا وثيقة ببراهين هندسية

" Ibn Khaldun remarks : A report has reached us, that some great scholars of the east have increased the number of cases beyond six, and have brought them to upwards of twenty, producing their accurate solutions together with geometrical demonstrations."

Page 8, *lime* 17.

See Leonardo's geometrical illustration of the three cases involving an affected square, as reported by Cossali (i. p. 2.), and hence by Hutton (Tracts, &c., ii. p. 198.)

Cardan, in the introduction of his *Ars magna*, distinctly refers to the demonstrations of the three cases given by our author, and distinguishes them from others which are his own. *At etiam demonstrationes, præter tres* Mahometis *et duas* Lodovici (Lewis Ferrari, Cardan's pupil), *omnes nostræ sunt.*—In another passage (page 20) he blames our author for having given the demonstration of only one solution of the case $cx^2 + a = bx$. *Nec admireris,* says he, *hanc secundam demonstrationem aliter quam a* MAHUMETE *explicatam, nam ille immutata figura magis ex re ostendit, sed tamem obscurius, nec nisi unam partem eamque pluribus.*

Page 17, *line* 11-13.

The words from في والاسدنا to وسدس السدس are written twice over in the manuscript.

Page 19, *line* 12.

" [جذر مال معلوم او اصم] " The root of a rational or irrational number." In the *Kholāset al Hisāb*, p. 128. 137. 369, the expression منطق (lit. audible) is used instead of معلوم, which stands in a more distinct opposition to اصم (lit. inaudible, surd). BAHA-EDDIN applies the same expressions also to fractions, calling منطق those for which there are peculiar expressions in Arabic, *e. g.* ثلث one-third, and اصم those which must be expressed periphrastically by means of the word جزء a part, *e. g.* ثلثة اجزاء من خمسة وعشرين three twenty-fifths. See *Kholāset al Hisāb,* p. 150.

Page 19, *line* 15.

The manuscript has مثلي ذلك المال . The context requires the insertion of جذر after مثلي, which I have added from conjecture.

Page 20, *line* 15. 17.

[ما يصيب الواحد] " What is proportionate to the unit,"

i. e. the quotient. This expression will be explained by BAHA-EDDIN's definition of division (*Kholáset al Hisáb*, p. 105). القسمة طلب عدد نسبته الى الواحد كنسبة المقسوم علیه المقسوم الى " Division is the finding a number which bears the same proportion to the unit, as the dividend bears to the divisor."

Page 21, line 17.

جذري [جذر . The MS. has

Page 24, line 6.

تمكنا لها صورة لا تحسن [An attempt at constructing a figure to illustrate the case of $[100 + x^2 - 20x] + [50 + 10x - 2x^2]$ has been made on the margin of the manuscript.

Page 30, line 10.

فخذ ما شئت [A marginal note in the manuscript defines this in the following manner. يعني اقسم العشرة كيف شئت اربعة حنطة وستة شعيرا وستة حنطة واربعة شعيرا او ثلثة حنطة وسبعة شعيرا اوعكس ذلك او كيف ما شئت فانه يصح العمل فيه حاشية من شرح المزيحفي " He means to say : divide the ten in any manner you like, taking four of wheat and six of barley, or four of barley and six of wheat, or three of wheat and seven of barley, or *vice versa*, or in any other way : for the solution will hold good in all these cases. (*Note from Al Mozaihafi's Commentary*)."

Page 42, line 8.

The manuscript has a marginal note to this passage,

2 c

from which it appears that the inconvenience attending the solution of this problem has already been felt by Arabic readers of the work.

Page 45, *line* 16.

This instance from MOHAMMED's work is quoted by CARDAN (*Ars Magna*, p. 22, edit. Basil.) As the passage is of some interest in ascertaining the identity of the present work with that considered as MOHAMMED's production by the early propagators of Algebra in Europe, I will here insert part of it. *Nunc autem*, says CARDAN, *subjungemus aliquas quæstiones, duas ex MAHUMETE, reliquas nostras.* Then follows *Quæstio I. Est numerus a cujus quadrato si abjeceris $\frac{1}{3}$ et $\frac{1}{4}$ ipsius quadrati, atque insuper 4, residuum autem in se duxeris, fiet productum æquale quadrato illius numeri et etiam 12. Pones itaque quadratum numeri incogniti quem quæris esse 1 rem, abjice $\frac{1}{3}$ et $\frac{1}{4}$ ejus, es insuper 4, fiet $\frac{5}{12}$ rei* m: *4, duc in se, fit $\frac{25}{144}$ quadrati* p: *16* m: *$3\frac{1}{3}$ rebus, et hoc est æquali uni rei et 12; abjice similia, fiet 1 res æqualis $\frac{25}{144}$ quadrati* p: *4* m: *$3\frac{1}{3}$ rebus*, &c.

The problem of the *Quæstio II.* is in the following terms, *Fuerunt duo duces quorum unusquisque divisit militibus suis aureos 48. Porro unus ex his habuit milites duos plus altero, et illi qui milites habuit duos minus contigit ut aureos quatuor plus singulis militibus daret ; quæritur quot unicuique milites fuerint.* In the present copy of MOHAMMED's algebra, no such instance occurs. Yet CAR-

DAN distinctly intimates that he derived it from our author, by introducing the problem which immediately follows it, with the words : *Nunc autem proponamus quæstiones nostras.*

Page 46, line 18.

The manuscript has the following marginal note to this passage : هذه المسئلة تعمل بالكعب و طريقه ان تاخذ مالا و تلقي ثلثه يبقي ثلثا مال تضرب ذلك في ثلثة اجذار فيكون كعبين يعدلان مالا فزده مرتين علي قدر المال يكون جذرين يعدلان درهما والجذر نصف المال والمال ربع اذا القيت ثلثه بقي سدس اذا ضربت ذلك في ثلثة اجذاره وهي درهم و نصف بلغ ذلك ربع درهم مثل المال كما ذكر " This instance may also be solved by means of a cube. The computation then is, that you take the square, and remove one-third from it; there remain two-thirds of a square. Multiply this by three roots; you find two cubes equal to one square. Extracting twice the square-root of this, it will be two roots equal to a dirhem. Accordingly one root is one-half, and the square one-fourth.* If you remove one-third of this, there remains one-sixth, and if you multiply this by three roots, that is by one dirhem and a half, it amounts to one-fourth of a dirhem, which is the square as he had stated."

---

* $[x^2 - \frac{1}{3}x^2] \times 3x = x^2$
$$2x^3 = x^2$$
$$2x = 1$$
$$x = \frac{1}{2}.$$

Page 50, line 2.

I am uncertain whether my translation of the definition which MOHAMMED gives of mensuration be correct. Though the diacritical points are partly wanting in the manuscript, there can, I believe, be no doubt as to the reading of the passage.

Page 51, line 12.

I have simply translated the words اهل الهندسة by "geometricians," though from the manner in which Mo-HAMMED here uses that expression it would appear that he took it in a more specific sense.

FIRUZABADI (Kamus, p. 814, ed. Calcutt.) says that the word *handasah* (الهندسة) is originally Persian, and that it signifies " the determining by measurement where canals for water shall be dug."

The Persians themselves assign yet another meaning to the word هندسه *hindisah*, as they pronounce it: they use it in the sense of decimal notation of numerals.*

It is a fact well known, and admitted by the Arabs

---

* هندسه بكسر اول وثالث و فتح سين بي نقطه بمعني
اندازه وشكل باشد وارقامي را نيزگويند كه در زير حروف
كلمات نويسند همچو ابجد هوز حطي
١٠٩٨ ٧٦٥ ٤٣٢١

"*Hindisah* is used in the sense of measurement and size; the same word is also applied to the signs which are written instead of the words (for numbers) as 1, 2, 3, 4, 5, 6, 7, 8, 9, 10." *Burhani Kati.*

themselves, that the decimal notation is a discovery for which they are indebted to the Hindus.*   At what time the communication took place, has, I believe, never yet been ascertained.   But it seems natural to suppose that it was at the same period, when, after the accession of the Abbaside dynasty to the caliphat, a most lively interest for mathematical and astronomical science first arose among the Arabs.   Not only the most important foreign works on these sciences were then translated into Arabic, but learned foreigners even lived at the court of Bagdad, and held conspicuous situations in those scientific establishments which the noble ardour of the caliphs had called forth. History has transmitted to us the names of several distinguished scholars, neither Arabs by birth nor Mohammedans by their profession, who were thus attached to the court of ALMANSUR and ALMAMUN; and we know from

---

* It is almost unnecessary to adduce further evidence in support of this remark. BAHA-EDDIN, after a few preliminary remarks on numbers, says وقد وضع لها حكماء الهند الارقام التسعة المشهورة "Learned Hindus have invented the well known nine figures for them." (*Kholáset al-Hisáb*, p. 16.)   In a treatise on arithmetic, entitled متن النزهة في علم الحساب which forms part of Sir W. OUSELEY's most valuable collection of Oriental manuscripts, the nine figures are simply called الاشكال الهندية.   See, on the subject generally, Professor VON BOHLEN's work, *Das alte Indien*, (Königsberg, 1830. 1831. 8.) vol. II. p. 224, and ALEXANDER VON HUMBOLDT's most interesting dissertation: *Ueber die bei verschiedenen Völkern üblichen Systeme von Zahlzeichen*, &c. (Berlin, 1829. 4.) page 24.

good authority, that Hindu mathematicians and astronomers were among their number.

If we presume that the Arabic word *handasah* might, as the Persian *hindisah*, be taken in the sense of decimal notation, the passage now before us will appear in an entirely new light. The اهل الهندسة, to whom our author ascribes two particular formulas for finding the circumference of a circle from its diameter, will then appear to be the Hindu Mathematicians who had brought the decimal notation with them;—and the اهل النّجوم منهم, to whom the second and most accurate of these methods is attributed, will be the Astronomers among these Hindu Mathematicians.

This conjecture is singularly supported by the curious fact, that the two methods here ascribed by Mohammed to the اهل الهندسة actually do occur in ancient Sanskrit mathematical works. The first formula, $p = \sqrt{10d^2}$, occurs in the *Vijaganita* (COLEBROOKE's translation, p. 308, 309.); the second, $p = \dfrac{d \times 62832}{20000}$, is reducible to $\dfrac{d \times 3927}{1250}$, the proportion given in the following stanza of BHASKARA's *Lilavati*:

व्यासे भनन्दाग्निहते विभक्ते
    खवाणसूर्यैः परिधिस्तु सूक्ष्मः ۱
द्वाविंशतिघ्ने विहृते च शैलैः
    स्थूलो ऽथ वा स्याद्व्यवहारयोग्यः ॥

" When the diameter of a circle is multiplied by three

thousand nine hundred and twenty-seven, and divided by twelve hundred and fifty, the quotient is the near circumference: or multiplied by twenty-two and divided by seven, it is the gross circumference adapted to practice."* (COLEBROOKE's translation, page 87. See FEIZI's Persian translation, p. 126, 127.)

The coincidence of $\frac{d \times 62832}{20000}$ with $\frac{d \times 3927}{1250}$ is so striking, and the formula is at the same time so accurate, that it seems extremely improbable that the Arabs should by mere accident have discovered the same proportion as the Hindus: particularly if we bear in mind, that the Arabs themselves do not seem to have troubled themselves much about finding an exact method.†

---

* The Sanskrit original of this passage affords an instance of the figurative method of ʻthe Hindus of expressing numbers by the names of objects of which a certain number is known: the expressions for the units and the lower ranks of numbers always preceding those for the higher ones. भ (lunar mansion) stands for 27; नन्द (treasure of Kuvera) for 9; and अग्नि (sacred fire) for 3: therefore भनन्दाग्नि = 3927. Again, ख (cypher) is 0; वाण (arrow of Kamadeva) stands for 5; सूर्य (the sun in the several months of the year), for 12: therefore खवाणसूर्य = 1250. For further examples, see *As. Res.* vol. XII. p. 281, ed. Calc., and the title-pages or conclusions of several of the Sanskrit works printed at Calcutta;—e. g. the *Sutras* of *Panini* and the *Siddhantakaumudi.*

† This would appear from the very manner in which our author introduces the several methods; but still more from the following marginal note of the manuscript to the present passage: وهو تقريب

Page 57, line 5-8.

The words between brackets are not in the manuscript: I have supplied the apparent hiatus from conjecture.

Page 61, line 4.

A triangle of the same proportion is used to illustrate this case in the *Lilavati* (FEIZI's Persian transl. p. 121. COLEBROOKE's transl. of the *Lilavati*, p. 71. and of the *Vijaganita*, p. 203.)

Page 65, line 12-14.

The words between brackets are in the manuscript written on the margin. I think that the context warrants me sufficiently for having received them into the text.

Page 66, line 5.

The words between brackets are not in the text, I give them merely from my own conjecture.

لا تحقيق ولا يقف احد علي حقيقة ذلك ولا يعلم دورها الا الله لان الخط ليس بمستقيم فيوقف علي حقيقته وانما قيل ذلك تقريب كما قيل في جذر الاصم انه تقريب لا تحقيق لان جذره لا يعلمه الا الله واحسن ما في هذه الاقوال ان تضرب القطر في ثلثة وسبع لانه اخف واسرع والله اعلم ۞ " This is an approximation, not the exact truth itself: nobody can ascertain the exact truth of this, and find the real circumference, except the Omniscient: for the line is not straight so that its exact length might be found. This is called an approximation, in the same manner as it is said of the square-roots of irrational numbers that they are an approximation, and not the exact truth: for God alone knows what the exact root is. The best method here given is, that you multiply the diameter by three and one-seventh: for it is the easiest and quickest. God knows best!"

Page 71, line 8, 9.

The author says, that the capital must be divided into 219320 parts: this I considered faulty, and altered it in my translation into 964080, to make it agree with the computation furnished in the note. But having recently had an opportunity of re-examining the Oxford manuscript, I perceive from the copious marginal notes appended to this passage, that even among the Arabian readers considerable variety of opinion must have existed as to the common denominator, by means of which the several shares of the capital in this case may be expressed.

One says : انظر لمال يكون لسدسه ربع والربعه ثلث وما بقي يتقسم علي ماية و خمسة وتسعين ولا يوجد ذلك في اقل من اربعة وعشرين فاضرب اربعة وعشرين في ماية وخمسة و تسعين يصح من ذلك اربعة الاف وستماية وثمانون ومنه يصح

"Find a number, one-sixth of which may be divided into fourths, and one-fourth of which may be divided into thirds; and what thus comes forth let be divisible by hundred and ninety-five. This you cannot accomplish with any number less than twenty-four. Multiply twenty-four by one hundred and ninety-five : you obtain four thousand six hundred and eighty, and this will answer the purpose."

Another :* وفي وجه اخر انك تجعل ماية وستة وخمسين

---

* The numbers in this and in part of the following scholium are in the MS. expressed by figures, which are never used in the text of the work.

سدس المال وتضربها في ٦ فيكون ٩٣٦ واذا استخرجت نصيب الابن وهو الثلث والربع وجدته ٥٤٦ و لا خمس لها فاضربها في ٥ يكون ٤٦٨٠ للام من ذلك ٤٢٥ و للزوج ٧٨٠ و للابن ٢٨٨ " Ac- و لصاحب الخمسين ١٤٩٢ والصاحب الربع ٦٩٥

cording to another method, you may take one hundred and fifty-six for the one-sixth of the capital. Multiply this by six; you find nine hundred and thirty-six. Taking from this the share of the son, which is one-third and one-fourth, you find it five hundred and forty-six. This is not divisible by five: therefore multiply the whole number of parts by five: it will then be four thousand six hundred and eighty. Of this the mother receives four hundred and twenty-five, the husband seven hundred and eighty, the son two hundred and eighty-eight (twelve hundred and eighty-eight?), the legatee, who is to receive the two-fifths, fourteen hundred and ninety-two, and the legatee to whom the one-fourth is bequeathed, six hundred and ninety-five."

Another: وفي [وجه] اخر يصح من تسعة الاف و ثلثماية و ستين ووجه العمل في ذلك ان [تقسم] الفريضة في اثني عشر للام سهمان وللزوج ثلثة وللابن سبعة فتضربها في ٢٠ لذكر الخمسي والربع فيكون مايتين واربعين فتاخذها سدسها اربعين للام والثلث جائز عليها وليس للاربعين ثلث فتضرب اصل المسئلة في ثلثة لذلك فيكون سبعماية وعشرين فتاخذ سدسها للام ماية وعشرين فيخرج من ذلك الثلث لاصحاب الوصايا وهو اربعون مقسوم على ثلثة عشر لا يصح فاضرب المسئلة في

١٣ يكون ٩٣٦٠ [MS. ٩٠٦٣] لما ذكرنا للام من ذلك

ثماني ماية و خمسون و للابن الفان و خمسماية و ستة و

سبعون و للزوج الف و خمسماية و ستون ولصاحب

الخمسين الفان و تسعماية واربعة وثمانون ولصاحب الربع

الف و ثلثماية و تسعون والله اعلم " According to another

method, the number of parts is nine thousand three hundred and sixty. The computation then is, that you divide the property left into twelve shares; of these the mother receives two, the husband three, and the son seven. This (number of parts) you multiply by twenty, since two-fifths and one-fourth are required by the statement. Thus you find two hundred and forty. Take the sixth of this, namely forty, for the mother. One-third out of this she must give up. Now, forty is not divisible by three. You accordingly multiply the whole number of parts by three, which makes them seven hundred and twenty. The one-sixth of this for the mother is one hundred and twenty. One-third of this, namely forty, goes to the legatees, and should be divided by thirteen; but as this is impossible, you multiply the whole number of parts by thirteen, which makes them nine thousand three hundred and sixty, as we said above. Of this the mother receives eight hundred and fifty, the son two thousand five hundred and seventy-six, the husband one thousand five hundred and sixty, the legatee to whom the two-fifths are bequeathed, two thousand nine hundred and eighty-four, and the legatee who is to receive one-fourth, one thousand three hundred and ninety."

www.ingramcontent.com/pod-product-compliance
Lightning Source LLC
Chambersburg PA
CBHW082036190526
45165CB00021B/3330